Ergebnisse der Mathematik und ihrer Grenzgebiete Band 85

Herausgegeben von P. R. Halmos P. J. Hilton R. Remmert B. Szőkefalvi-Nagy

Unter Mitwirkung von L. V. Ahlfors R. Baer F. L. Bauer A. Dold J. L. Doob S. Eilenberg K. W. Gruenberg M. Kneser G. H. Müller M. M. Postnikov B. Segre E. Sperner

Geschäftsführender Herausgeber P. J. Hilton

Wu Yi Hsiang

Cohomology Theory of Topological Transformation Groups

Springer-Verlag
Berlin Heidelberg New York 1975

Wu Yi Hsiang
Department of Mathematics, University of California, Berkeley

AMS Subject Classification (1970): 57 Exx

ISBN-13: 978-3-642-66054-2 e-ISBN-13: 978-3-642-66052-8
DOI: 10.1007/978-3-642-66052-8

Library of Congress Cataloging in Publication Data. Hsiang, Wu Yi, 1937 –. Cohomology theory of topological transformation groups. (Ergebnisse der Mathematik und ihrer Grenzgebiete; Bd. 85). Bibliography: p. Includes index. 1. Transformation groups. 2. Topological groups. 3. Homology theory. I. Title. II. Series. QA613.7.H85. 514'.2. 75-5530.

Softcover reprint of the hardcover 1st edition 1975

Introduction

Historically, applications of algebraic topology to the study of topological transformation groups were originated in the work of L. E. J. Brouwer on periodic transformations and, a little later, in the beautiful fixed point theorem of P. A. Smith for *prime periodic maps* on homology spheres. Upon comparing the fixed point theorem of Smith with its predecessors, the fixed point theorems of Brouwer and Lefschetz, one finds that it is possible, at least for the case of homology spheres, to upgrade the conclusion of mere existence (or non-existence) to the actual *determination of the homology type of the fixed point set, if the map is assumed to be prime periodic.* The pioneer result of P. A. Smith clearly suggests a fruitful general direction of studying topological transformation groups in the framework of algebraic topology. Naturally, the immediate problems following the Smith fixed point theorem are to generalize it both in the direction of replacing the homology spheres by spaces of more general topological types and in the direction of replacing the group $\mathbb{Z}_p$ by more general compact groups. It is usually rather straightforward to deduce similar fixed point theorems for actions of *p-primary groups* (or extensions of torus groups by p-primary groups) directly from the corresponding fixed point theorems for actions of the group $\mathbb{Z}_p$. However, various efforts to extend such fixed point theorems *beyond p-primary groups* (or extensions of torus by p-primary groups) all eventually wound up with puzzling counter-examples [C6, C8, F2, H5]. On the other hand, if the group is an elementary p-group, (ie., $\mathbb{Z}_p^r$ or T^r if $p=0$), then a far-reaching generalization of the Smith fixed point theorem, valid for all finite dimensional, locally compact spaces, can be formulated and proved in the framework of equivariant cohomology. (cf. Theorem (IV.1), § 1, Ch. IV).

The basic setting for our approach using the cohomology theory in compact topological transformation groups is the following *equivariant cohomology theory* introduced by A. Borel [B 5]. Let G be a compact Lie group and let X be a given G-space. Then the *equivariant cohomology of the G-space X* is defined to be the *ordinary cohomology of the total space* X_G of the universal bundle, $X \to X_G \to B_G$, with the given G-space as the typical fibre. The reasons for adopting such an equivariant cohomology theory in terms of the universal bundle construction are roughly the following:

(i) Intuitively and heuristically, the complexity of the G-action on X will be reflected in the complexity of the associated universal bundle, $X \to X_G \to B_G$; and the classical characteristic class theory clearly demonstrates that cohomology theory can then be used to detect the complexity of the bundle, which, in turn, also detects the complexity of the G-action on X itself. Therefore, the above definition of equivariant cohomology simply formalizes and also generalizes the classical characteristic class theory to the study of the topology of general fibre bundles.

(ii) From a technical standpoint, the above equivariant cohomology theory naturally and successfully brings together the modern theories of fibre bundles, spectral sequences and sheaves in a nice convenient way. Hence, it not only possesses all the convenient formal properties that one expects, but also is effectively computable.

Basic properties of this equivariant cohomology theory as well as some fundamental general theorems such as the localization theorem of Borel-Atiyah-Segal type are formulated and proved in Chapter III.

In Chapter IV, we shall proceed to investigate the relationship between the *geometric structures* of a given G-space X and the *algebraic structures* of its equivariant cohomology $H_G^*(X)$. From the viewpoint of transformation groups, those structures which are usually summarized as the *orbit structure* are certainly the most important geometric structures of a given G-space X. Hence, it is almost imperative that one should investigate how much of the orbit structure of X can actually be determined from the algebraic structure of $H_G^*(X)$. Examples of specific problems in this area are: How much of the cohomology structure of the fixed point set, $H^*(F)$, is determined by the algebraic structure of $H_G^*(X)$? Is it possible to give a criterion for the existence of fixed points purely in terms of $H_G^*(X)$? Suppose $F=F(G,X)=\emptyset$ (is empty), how does one determine the set of *maximal isotropy subgroups* from $H_G^*(X)$? *In the special case of elementary p-groups*, we shall formulate various commutative algebraic invariants of $H_G^*(X)$, which are, then, proved to be intimately related to the orbit structure of the G-space X. Notice that there are general *counter-examples* for almost all *non-p-primary groups* which clearly indicate the *non-existence* of a general relationship between the orbit structure of X and the algebraic structure of $H_G^*(X)$. Such a sharp contrast of behaviors between transformations of *elementary p-groups* and transformations of *non-p-primary groups* is probably one of the most profound as well as fascinating facts in the cohomology theory of transformation groups. In retrospect, this also explains why *torus groups* play such a central rôle in the representation theory of compact connected Lie groups, which, after all, is concerned with the special case of *linear* transformation groups.

Methodologically, one of the central themes in the approach of this book is that the cohomology theory of *topological* transformation groups can be developed roughly along the same lines as the classical *linear* representation theory of compact connected Lie groups. A concise exposition of the theory of compact Lie groups and their linear representations is given in Chapter II. In order to present the theory of *linear* transformation groups as a prototype of cohomology theory of *topological* transformation groups, we purposely adopt a rather geometric approach, in which the orbit structure of the adjoint action plays the central rôle. Moreover, it will be clear from such an exposition that the following two basic theorems constitute the foundation of *linear* representation theory of compact connected Lie groups:

(i) *Structural splitting theorem for linear tori actions:* every complex *linear* representiation of a torus group always splits into the sum of one-dimensional representations.

(ii) *Maximal tori theorem of É. Cartan:* the set of maximal tori forms a single conjugacy class and $G=\bigcup\{gTg^{-1}; g\in G\}$.

The first result *classifies linear* tori actions in terms of an extremely simple invariant called the *weight system* and the second result reduces the classification of *linear* actions of a compact connected Lie group G to the restricted actions of its maximal tori. Correspondingly, in the setting of the cohomology theory of *topological* transformation groups, the above structural splitting theorem for *linear* tori actions can be generalized into various structural splitting theorems of the equivariant cohomology (cf. Chapter IV), which can be considered as the generalized splitting principle in the geometric theory of generalized charateristic classes. Similar to the *linear* case, one may also combine the structural splitting theorems with the maximal tori theorem to define a *(geometric) weight system* for *topological* transformation groups. Such a program is carried out explicitly for the special cases of acyclic manifolds and cohomology spheres in Chapter V; and for cohomology projective spaces in Chapter VI. Although the geometric weight systems are no longer "complete invariants" for topological transformation groups, they nevertheless determine the cohomology aspects of orbit structures of the restricted actions of maximal tori and, hence, also the orbit structure of the original G-action to a great extent.

In Chapter VII, we apply the cohomology method to study transformation groups on compact homogeneous spaces. Due to the fact that compact homogeneous spaces encompass great varieties of topological types and that the study of transformation groups on them is just getting started, there is an abundance of natural problems in this area, but, so far, only a small number of testing cases have been successfully settled and most of them are as yet unpublished. Therefore, results in this chapter are rather incomplete, and they should be considered only as beginning testing cases that serve to indicate the existence of interesting problems and deep results. In a paper soon to be published, I hope to give a more systematic account of the applications of the cohomology method to the study of transformation groups on compact homogeneous spaces.

This book is based on a course given at the University of California, Berkeley. I am indebted to the participants of that course, especially to Dr. T. Skjelbred who helped to prepare a preliminary draft of Chapters III and IV.

Berkeley, in Spring 1975

Wu Yi Hsiang

Table of Contents

Chapter I. Generalities on Compact Lie Groups and G-Spaces

This chapter will briefly review the general facts about compact topological groups, fibre bundles, topological G-spaces and compact Lie groups that are necessary for the subsequent development. Basic concepts and definitions will be adequately explained; and proofs of some fundamental theorems will also be included whenever short clear cut proofs are available.

§ 1. General Properties of Compact Topological Groups

Naturally, a topological group G consists of both a *topological structure* and a *group structure* which are *compatible* in the sense that the group structure is *continuous* with respect to the topological structure. More precisely, the multiplication and inversion mappings are both continuous: $G \times G \to G$, $(g_1, g_2) \mapsto g_1 \cdot g_2$; $G \to G$, $g \mapsto g^{-1}$. Similarly, a Lie group (or rather a differentiable group) G consists of both a *differentiable structure* and a *group structure* which are *compatible* in the sense that the multiplication and inversion mappings are both differentiable. For many problems, the subclass of compact topological groups, or more specifically, compact Lie groups plays an important rôle. In this section, we shall summarize the basic properties of compact topological groups:

(A) *Averaging and Haar Measure*

Obviously, a finite group G (with discrete topology) is a rather special example of compact topological group. Suppose $\varphi: G \to Gl(V)$ is a given representation which represents G as a group of linear transformations on a vector space V. The following well known "averaging method" is a natural, simple-minded way to show the existence of an invariant inner product on V (with respect to the action of G). Given an arbitrary inner product $\langle x, y \rangle$ on V, it is clear that the following "*averaged inner product*" (x, y):

$$(x, y) = \frac{1}{\operatorname{ord}(G)} \sum_{g_i \in G} \langle g_i x, g_i y \rangle$$

is an invariant inner product on V with respect to the action of G. And this is also the only effective way, so far, to prove the *complete reducibility* of representations of finite groups.

Next let us consider the group of unit complex numbers $S^1=\{z\in\mathbb{C};\ |z|=1\}$. Topologically, it is a circle and it is one of the simplest example of compact topological group with an infinite number of elements. If we parametrize the circle group in the usual way, i.e. $S^1=\{e^{2\pi i\theta};\ 0\leqslant\theta\leqslant 1\}$, and f is a continuous function on S^1, then the integration of f

$$\int_0^1 f(\theta)\,d\theta$$

is clearly a generalized "*average value of* f". Similarly, to any linear representation $\varphi: S^1\to Gl(V)$ and a given inner product $\langle x,y\rangle$ on V, the following "(generalized) *averaged inner product*" is again invariant with respect to the given action of S^1 on V.

$$(x,y)=\int_0^1 \langle e^{2\pi i\theta}\cdot x,\ e^{2\pi i\theta}\cdot y\rangle\,d\theta\,.$$

As for a general compact group G, it is natural to blend together the above two kind of "averagings". Namely, for a finite subset $A=\{a_j\}\subseteq G$ and a continuous function $f(g)$, one defines the *averaging* of f over A to be

$$(A\cdot F)(g)=\frac{1}{n}\sum_{j=1}^{n} f(a_j\cdot g)\,.$$

Heuristically, it is reasonable to expect that $\{A\cdot f\}$ will tend to a constant function, $I(f)$, as a limit when A becomes more and more dense in G. This is exactly the idea of Von Neumann in defining the (invariant) Haar-integral on a compact group G. We state the result as the following theorem and refer to Pontrjgin's book "Topological Groups" for a detail proof.

Theorem I.1 (Existence and uniqueness of Haar integral). *Let G be a given compact topological group and $C(G)$ be the space of real-valued continuous functions of G. Then there exists a unique continuous linear functional $I: C(G)\to\mathbb{R}$ satisfying*

(i) *(left invariant): $I(f_a(g))=I(f)$ for all $a\in G$, where $f_a(g)=f(ag)$.*

(ii) *(positive and normalized): $f\geqslant 0\Rightarrow I(f)\geqslant 0$, and $I(1)=1$.*

The above invariant linear functional I can also be considered as the *integral* with respect to the invariant measure—the Haar measure—on G with total volume 1, i.e., $I(f)=\int_G f(g)\,dg$.

Corollary (I.1.1). *Every complex (resp. real) representation of compact group G is equivalent to a unitary (resp. orthogonal) represenation.*

Corollary (I.1.2). *Every complex (resp. real) representation φ of a compact group G is completely reducible. Hence it decomposes uniquely into the direct sum of irreducible representations.*

(B) *Schur Lemma and Schur Orthogonality Relations*

The Schur lemma is simply a neat reformulation of irreducibility purely in terms of operators; however, it is extremely useful and is of fundamental importance.

Schur Lemma. *Let S_1, S_2 be irreducible subsets of linear transformations on V_1, V_2 respectively and $A: V_1 \to V_2$ be a linear mapping such that $A \cdot S_1 = S_2 \cdot A$ (set-wise). Then either $A=0$ or A is invertible.*

Proof. It follows easily from $A \cdot S_1 = S_2 \cdot A$ that $\operatorname{Ker} A$ and $\operatorname{Im}(A)$ are invariant subspaces of S_1, S_2 respectively. Hence, the irreducibility of S_1, S_2 implies that either $\operatorname{Ker} A = V_1$, i.e., $A=0$; or $\operatorname{Ker} A = \{0\}$ and $\operatorname{Im}(A) = V_2$, i.e., A is invertible. □

The following special form of Schur's lemma is often useful and deserves special attention.

Special Form of Schur's Lemma. *In the special case that $S_1 = S_2$ (then of course $V_1 = V_2$), and the field F is algebraically closed then there exists $\lambda_0 \in F$ such that $A = \lambda_0 \cdot I$.*

Proof. $A \cdot S_1 = S_1 \cdot A \Rightarrow (A - \lambda I) \cdot S_1 = S_1 \cdot (A - \lambda I)$ for any $\lambda \in F$. On the other hand, since F is assumed to be algebraically closed, there exists eigen-value $\lambda_0 \in F$ of A, i.e., $(A - \lambda_0 I)$ is non-invertible. Hence, by the above lemma $(A - \lambda_0 I) = 0$, i.e., $A = \lambda_0 I$. □

As a direct consequence, we have the following simple but fundamental theorem for representation of compact commutative groups.

Theorem (I.2). *Every irreducible complex representation φ of a compact commutative group G is one-dimensional.*

Proof. It follows from the *commutativity* of G and the *irreducibility* of $\varphi(G)$ that for every $g \in G$, $\varphi(g) = \lambda(g) \cdot I$, for some $\lambda(g) \in \mathbb{C}$.

Thus, $\varphi(G) \subseteq \{\lambda \cdot I\}$ = the set of scalar multiples. Hence, every subspace is invariant and $\varphi(G)$ is irreducible only when $\dim \varphi = 1$. □

Corollary (I.2.1) (Classification of irreducible representation of torus group). *The set of equivalence classes of irreducible complex representations of a torus group of rank k, $T^k \cong \mathbb{R}^k/\mathbb{Z}^k$, is naturally in $1-1$ correspondence with the set of integral linear functional of $\mathbb{R}^k$. Hence, it is also naturally in $1-1$ correspondence with elements of $H^1(T^k, \mathbb{Z})$,* i.e.,

$$\{\varphi: T^k \to S^1\} \leftrightarrow \{\varphi^*: H^1(S^1, \mathbb{Z}) \to H^1(T^k, \mathbb{Z})\} \leftrightarrow \{\varphi^*(i) \in H^1(T^k, \mathbb{Z})\}$$

where $i \in H^1(S^1, \mathbb{Z})$ is the fundamental class with positive orientation.

Definition. With respect to a chosen basis, let a unitary representation $\varphi: G \to U(n)$ be given by its matrix form $\varphi(g) = (\varphi_{ij}(g))$. Then $\varphi_{ij}(g)$ are called the *representation functions* of φ.

Example. In the well known classical case of $G = S^1$, the representation function corresponding to the irreducible representation $\varphi: S^1 \to S^1$ with winding number

n is $e^{2\pi in\theta}$, and it is the fundamental fact of Fourier analysis that the set of such representation functions, $\{e^{2\pi in\theta}; n\in\mathbb{Z}\}$, of all irreducible representations of S^1 forms an orthonormal basis of $L^2(S^1)$.

Combining the Schur lemma with the invariant integration, one has the following orthogonality relations for representation functions of a general compact group G.

Theorem (I.3) (Schur orthogonality relations). *Let G be a given compact group and φ,ψ are two non-equivalent, irreducible, complex representations of G. Then*

(i) $\int_G \varphi_{ij}(g)\overline{\psi_{kl}(g)}\,dg=\langle\varphi_{ij}(g),\psi_{kl}(g)\rangle=0$,

(ii) $\int_G \varphi_{ij}(g)\cdot\overline{\varphi_{kl}(g)\,dg}=\langle\varphi_{ij},\varphi_{kl}\rangle=\dfrac{1}{\dim\varphi}\cdot\delta_{ik}\cdot\delta_{jl}$.

Proof. Let V,W be the representation space of ψ,φ and $\mathscr{L}(V,W)\cong V^*\otimes W$ be the space of all linear mappings of V into W. Define the induced representation on $\mathscr{L}(V,W)$ by

$$g\cdot A=\varphi(g)\cdot A\cdot\psi(g)^{-1}=\varphi(g)A\cdot\overline{\psi(g)}^t,\qquad A\in\mathscr{L}(V,W).$$

Then, it is clear that $\tilde{A}=\int_G(g\cdot A)\,dg$ is an invariant element, i.e., $g\cdot\tilde{A}=\varphi(g)\cdot\tilde{A}\cdot\psi(g)^{-1}=\tilde{A}$ for all $g\in G$. Hence, it follows from Schur lemma that $\tilde{A}$ is either 0 or invertible. But φ,ψ are non-equivalent, $\tilde{A}$ must be 0 for all $A\in\mathscr{L}(V,W)$. If we write down the equation $\tilde{E}_{jk}=0$ for the usual basis E_{jk} of $\mathscr{L}(V,W)$ in matrix form, we get

$$\tilde{E}_{jk}=\int_G\varphi(g)\cdot E_{jk}\cdot\overline{\psi(g)}^t\,dg=\int_G\varphi_{ij}(g)\overline{\psi_{kl}(g)}\,dg=0.$$

Similarly, it follows from the special form of Schur's lemma that

$$\tilde{B}=\int_G(g\cdot B)\,dg=\int_G\varphi(g)B\varphi(g)^{-1}\,dg=\lambda_B\cdot I\quad\text{for}\quad B\in\mathscr{L}(W,W)$$

and suitable $\lambda_B\in\mathbb{C}$. On the other hand, the invariance of trace under conjugation enables us to compute λ_B as follows:

$$\dim\varphi\cdot\lambda_B=\operatorname{tr}(\tilde{B})=\int_G\operatorname{tr}(gB)\,dg=\int_G\operatorname{tr}(B)\,dg=\operatorname{tr}(B).$$

Hence, again, we have

$$\tilde{E}_{ik}=\int_G\varphi(g)\cdot E_{ik}\varphi(g)^{-1}\,dg=\frac{\delta_{ik}}{\dim\varphi}\cdot I\Rightarrow\int_G\varphi_{ij}(g)\overline{\varphi_{kl}(g)}\,dg=\frac{\delta_{ik}\cdot\delta_{jl}}{\dim\varphi}.\quad\square$$

(C) *Characters: A Complete Invariant of Linear Representations*

Definition. $\chi_\varphi(g)\overset{\text{def}}{=}\operatorname{tr}(\varphi(g))=\sum_{i=1}^n\varphi_{ii}(g)$ is called the *character* (function) of the complex representation φ of G.

Observations. (i) Since $\operatorname{tr}(\varphi(g))=$ the sum of eigen-values of $\varphi(g), \chi_\varphi(g)$ may be considered as a *linear-type* invariant of $\varphi(g)$,

(ii) $\chi_{\varphi+\psi}(g)=\operatorname{tr}((\varphi+\psi)(g))=\operatorname{tr}(\varphi(g))+\operatorname{tr}(\psi(g))=\chi_\varphi(g)+\chi_\psi(g)$,

(iii) $\chi_{\varphi\otimes\psi}(g)=\operatorname{tr}(\varphi\otimes\psi(g))=\operatorname{tr}(\varphi(g))\cdot\operatorname{tr}(\psi(g))=\chi_\varphi(g)\cdot\chi_\psi(g)$,

(iv) $\chi_\varphi(g_1 g g_1^{-1})=\operatorname{tr}(\varphi(g_1)\varphi\cdot(g)\varphi(g_1)^{-1})=\operatorname{tr}(\varphi(g))=\chi_\varphi(g)$.

One of the most fundamental consequences of Theorem (I.3) is that the character function $\chi_\varphi(g)$ *completely classifies* the representation.

Theorem (I.3′). (i) *The characters of two non-equivalent, irreducible complex representations φ and ψ are orthogonal to each other,* $\langle\chi_\varphi,\chi_\psi\rangle=0$.

(ii) *A complex representation φ is irreducible if and only if the norm of χ_φ is* 1, *i.e.*,

$$\|\chi_\varphi\|=\langle\chi_\varphi,\chi_\varphi\rangle=\int_G \chi_\varphi\cdot\overline{\chi}_\varphi\, dg=1\,.$$

(iii) *Two complex representation φ_1,φ_2 (not necessary irreducible) are equivalent if and only if*

$$\chi_{\varphi_1}(g)=\chi_{\varphi_2}(g) \quad \textit{for all} \quad g\in G\,.$$

Proof. Follows immediately from Theorem (I.3) and complete reducibility.

Remarks. (i) Since the character function χ_φ has the same value on conjugate elements, χ_φ can be viewed as a function on the set of conjugacy classes G/Ad. Furthermore if $H\subseteq G$ is a closed subgroup of G which intersects *every* conjugacy class of G, then

$$\varphi_1\sim\varphi_2\Leftrightarrow\chi_{\varphi_1}=\chi_{\varphi_2}\Leftrightarrow\chi_{\varphi_1}|H=\chi_{\varphi_2}|H\Leftrightarrow\chi_{\varphi_1|H}=\chi_{\varphi_2|H}\Leftrightarrow\varphi_1|H\sim\varphi_2|H\,.$$

(ii) The compactness of G implies that $\varphi(g)$ are unitary [Cor. (1.1)] and hence diagonalizable. Therefore the equivalence class of $\varphi(g)$ is completely determined by the basic symmetric forms $\{\sigma_i\}$ of its eigen-values $\{\lambda_j\}$. On the other hand, since the set $\varphi(G)$ is closed under multiplication, the character $\chi_\varphi(g)=\operatorname{tr}\varphi(g)$ $=\sigma_1(\lambda_j)$ not only gives us $\sigma_1(\lambda_j)$, but also $\chi_\varphi(g^k)=\operatorname{tr}(\varphi(g)^k)=\sum_{j=1}^n\lambda_j^k$, which in turn gives back $\{\sigma_j, j=1,\ldots,n\}$. Roughly, this is the basic reason why such a simple, linear-type invariant as the character already suffices to classify them. One also sees that both the *compactness* and the *group structure* play a vital rôle.

(D) *Completeness Theorem of Peter-Weyl*

Theorem (I.4) (Peter-Weyl). *For a given compact topological group G, let Γ be a complete collection of non-equivalent, irreducible complex representation of G. Then the linear combinations of $\{\varphi_{ij}(g), \varphi\in\Gamma\}$ are uniformly dense in the space of continuous functions on G, $C(G)$. Or equivalently, $\{\varphi_{ij}(g), \varphi\in\Gamma\}$ forms a basis in the Hilbert space of L^2-functions on G, $L^2(G)$.*

We refer the reader to Pontrjgin's book for a standard proof of this theorem.

Definition. A representation φ is called *faithful* if it is an isomorphism into.

Corollary (I.4.1). *Every compact Lie group has a faithful representation.*

Proof. Since $\{\varphi_{ij}(g), \varphi\in\Gamma\}$ span $C(G)$, they clearly seperate points. Hence, it is a direct consequences of Peter-Weyl theorem that

$$\bigcap\{\ker(\varphi);\ \varphi\in\Gamma\}=\{\mathrm{id}\} = \text{the identity subgroup}.$$

Therefore

$$\bigcap\{\ker\varphi-U);\ \varphi\in\Gamma\}=\emptyset$$

for any neighborhood U of the identity, and it follows from the compactness of G (or rather the compactness of $(G-U)$), that there exist a finite subcollection $\varphi_1,\ldots,\varphi_k\in\Gamma$ with $\bigcap_{i=1}^k\{\ker\varphi_i-U\}=\emptyset$.

$$\bigcap_{i=1}^k\{\ker\varphi_i\}=\ker(\varphi_1+\cdots+\varphi_k)\subseteq U\,.$$

On the other hand, it is easy to show that for a small neighborhood U in a *Lie group*, the only subgroup contained in U is the identity subgroup. Hence

$$\ker(\varphi_1+\cdots+\varphi_k)\subseteq U\Rightarrow\ker(\varphi_1+\cdots+\varphi_k)=\{\mathrm{id}\}$$

i.e., $\varphi_1+\cdots+\varphi_k$ is a faithful representation of G. □

Remark. (i) Since every closed subgroup of a Lie group is also a Lie group, a compact topological group has a faithful representation when and only when it is a Lie group.

(ii) For a compact group G with a given faithful representation, it is not difficult to deduce the Peter-Weyl theorem of G by using the Stone-Weistrass approximation theorem. Hence, for compact Lie groups, the Peter-Weyl theorem is equivalent to the existence of faithful representations.

§ 2. Generalities of Fibre Bundles and Free G-Spaces

(A) *The Concept of G-spaces and Fibre Bundles*

Definition. A *topological transformation* group consists of a topological space X, a topological group G and a transformation of G on X (i.e., a map $G\times X\to X$; $(g,x)\mapsto g\cdot x$ with $e\cdot x=x$, $g_1\cdot(g_2\cdot x)=(g_1\cdot g_2)\cdot x$) which is *continuous* in the sense that the map $G\times X\to X$; $(g,x)\mapsto g\cdot x$ is continuous. Sometimes, we also call such a structure a *topological G-action on X*, or simply a *topological G-space*.

Examples. (i) The *linear* transformation groups of

$$\mathrm{GL}(n,\mathbb{R}) \text{ on } \mathbb{R}^n, \text{ or } \mathrm{GL}(n,\mathbb{C}) \text{ on } \mathbb{C}^n,$$

or their restrictions via some linear representations:

$$G\xrightarrow{\varphi}\mathrm{GL}(n,\mathbb{R})\ \ (\text{or } \mathrm{GL}(n,\mathbb{C}))$$

are obviously topological transformation groups.

(ii) The isometry group of a Riemannian manifold, (equipped with compact open topology) is a topological transformation group.

Intuitively, a fibre bundle consists of a projection $p: B \to X$ from the *bundle space* B onto the *base space* X which is locally a product but globally twisted. One of the simplest such example is the Möbius band which is a twisted product of the circle S^1 and the interval I with an orientation reversion $\mathbb{Z}_2$-twist on I. Technically and theoretically, it is important to pin down precisely what are the permissable twistings. This is exactly where the so called structural group G and the G-space structure on the fibre Y enter into the formal definition of fibre bundles.

Definition. A fibre bundle consists of a bundle space B, a base space X, a fibre Y with a *given* topological G-action and a projection $p: B \to X$ together with compatible local product structures. Namely, there exists a family of open coverings $\{V_j\}$ of X and local product structures, $\phi_j: V_j \times Y \to p^{-1}(V_j) \subseteq B$ such that the *twistings* between two overlabing local product structures are provided by the *given* G-action on Y, i.e.,

$$\begin{array}{ccccccc}
V_i \times Y & \supseteq & (V_i \cap V_j) \times Y & \xrightarrow{\phi_j^{-1}\phi_i} & (V_i \cap V_j) \times Y & \subseteq & V_j \times Y \\
\downarrow{\scriptstyle \phi_i} & & \searrow{\scriptstyle \phi_i} & & \swarrow{\scriptstyle \phi_j} & & \downarrow{\scriptstyle \phi_j} \\
p^{-1}(V_i) & \supseteq & & p^{-1}(V_i \cap V_j) & & \subseteq & p^{-1}(V_j)
\end{array}$$

where $\phi_j^{-1}\phi_i(x,y) = g_{ji}(x) \cdot y$, $x \in (V_i \cap V_j)$, $y \in Y$ and $g_{ji}: V_i \cap V_j \to G$ is continuous.

Remark. The above definition involves a choice of *coordinate* neighborhoods $\{V_i\}$ and local product structures $\{\phi_i\}$, hence it is not intrinsic and there is an obvious kind of equivalence relation among two such structures. We refer to § 2,3 of Steenrod's book [S 11] for a detail discussion on this matter.

(B) *Principal Bundle and Principal Map*

Definition (of bundle map). Let B, B' be two fibre bundles with the same G-space Y as fibre. Then a continuous map $h: B \to B'$ is called a *bundle map* if

(i) h carries each fibre Y_x of B homeomorphically onto a fibre $Y_{x'}$ of B', thus inducing a map $\bar{h}: X \to X'$ with $p'h = \bar{h}p$,

(ii) it is compatible with the twisting in the sense that

$$\begin{array}{ccc}
V_i \cap \bar{h}^{-1}(U_j) \times Y & \xrightarrow{\phi_j'^{-1} h \phi_i} & U_j \times Y \\
\downarrow{\scriptstyle \phi_i} & & \downarrow{\scriptstyle \phi_j'} \\
p^{-1}(V_i \cap \bar{h}^{-1}(U_j)) & \xrightarrow{h} & p'^{-1}(U_j)
\end{array}$$

where $\phi_j'^{-1} h \phi_i(x,y) = \bar{g}_{ji}(x) \cdot y$, and $\bar{g}_{ji}: V_i \cap \bar{h}^{-1}(U_j) \to G$ is continuous.

Remarks. (i) In the simplest case of the bundle $Y \to \{pt\}$ over a point, the operation of an element $g \in G$, $Y \to Y: y \mapsto g \cdot y$, is clearly a bundle map and conversely, every bundle map $Y \to Y$ also concides with the operation of some element of G. Hence, there is a bijection between the set of all bundle maps of Y onto itself and $\bar{G} = G/K_0$, where $K_0 = \{g \in G; g \cdot y = y \text{ for all } y \in Y\}$ is the ineffective kernel.

(ii) It is obvious that the composition of bundle maps is a bundle map.

Principal bundle. Let $B \xrightarrow{p} X$ be a given fibre bundle with a given G-space Y as typical fibre. Naturally, the set $\bar{B}$ of all *bundle maps* of the simplest bundle $Y \to \{pt\}$ into $B \xrightarrow{p} X$ constitutes an important structural invariant. As usual, we shall first try to equip $\bar{B}$ with as many natural structures as possible. Since the composition of two bundle maps is a bundle map, there is a natural (right) action of $\bar{G}$ on $\bar{B}: \bar{B} \times \bar{G} \to \bar{B}$, $(\bar{b}, \bar{g}) \mapsto b \circ g$. Moreover, there is a projection $\bar{p}: \bar{B} \to X$ by assigning each bundle map $\bar{b}$ to the induced image in the base, i.e., $\bar{p}(\bar{b}) = p\bar{b}(y) \in X$, and it is not difficult to show that $X \cong \bar{B}/\bar{G}$ (bijection).

Definition (Principal bundle). The *principal bundle* associated to a bundle $B \xrightarrow{p} X$ is the set $\bar{B}$ of all bundle maps of $Y \to \{pt\}$ into $B \xrightarrow{p} X$, equipped with the above right $\bar{G}$-action and the projection $\bar{B} \xrightarrow{\bar{p}} X \cong \bar{B}/\bar{G}$.

Remark. It follows easily from the *local product* structure of $B \xrightarrow{p} X$ that $\bar{B} \xrightarrow{\bar{p}} X$ is also locally a product. Hence, it is easy to equip $\bar{B}$ with a unique suitable topology so that the bijection $X \cong \bar{B}/\bar{G}$ becomes a homeomorphism.

Definition. The evaluation map $\bar{B} \times Y \xrightarrow{P} B$, $(\bar{b}, y) \mapsto \bar{b}(y)$ is called the *principal map*.

Remark. (i) In the case that G acts effectively on Y, i.e., $K_0 = \{\text{id}\}$, then $\bar{G} = G$ and $\bar{B}$ is a free (right) G-space and the above principal map $P: \bar{B} \times Y \to B$ simply identifies B with the quotient space $\bar{B} \times_G Y = \bar{B} \times Y/\text{"}\sim\text{"}$, where $(\bar{b}, y) \sim (\bar{b} \cdot g^{-1}, g \cdot y)$. In general $\bar{B}$ is a free (right) $\bar{G}$-space which can also be considered as a (right) G-space naturally. Then again, the map P identifies $\bar{B} \times_G Y \cong B$.

(ii) The principal map P indicates how to recover the bundle $B \xrightarrow{p} X$ from the principal bundle $\bar{B}$ and the given G-space Y. Hence, for a fixed G-space Y, the principal bundle $\bar{B}$ constitutes a *complete invariant.*

(C) *Intrinsic Definition of Bundles and Gleason Lemma*

$(X \times G)$ has an obvious (right) G-action given by $(x, g) \cdot g_1 = (x, g \cdot g_1)$. A free (right) G-space E is called locally trivial if to every point $x \in E/G$ (orbit space of E) there exist a neighborhood U, such that $p^{-1}(U)$ is isomorphic to the obvious (right) G-space $U \times G$. In view of the principal bundle and principal map, it is natural to give the following intrinsic definition of fibre bundle:

Definition. A fibre bundle consists of a locally trivial free (right) G-space $\bar{B}$ (called the principal bundle) and a (left) G-space Y (called the fibre). Then the orbit space $X = \bar{B}/G$ is called the base space; the quotient space $B = \bar{B} \times_G Y = \bar{B} \times Y/\text{"}\sim\text{"}$, where $(\bar{b}, y) \sim (\bar{b} g^{-1}, g y)$ is called the bundle space (or total space) and the following map p is called the projection:

$$\begin{array}{ccc} \bar{B}\times Y & \xrightarrow{p_1} & \bar{B} \\ \Big\downarrow \text{“}\sim\text{”} & & \Big\downarrow \bar{p} \\ B=\bar{B}\times_G Y & \xrightarrow{p} & \bar{B}/G=X \end{array}$$

where p_1 is the projection onto the first factor and vertical map are projections onto respective "orbit spaces".

In many useful cases, the structural group G is either automatically a compact Lie group or can be reduced to a compact Lie group (cf. (E)). In those cases, the following lemma of Gleason adds a convenient bonus to the above definition, namely, the local triviality condition holds *automatically* for *any* free G-space when G is a compact Lie group.

Gleason Lemma. *If G is a compact Lie group, then any free G-space is locally trivial.*

(We shall prove a more general form of this lemma in § 3.)

(D) *Homotopy Lifting Property, Induced Bundles and Classifying Bundles*

For the purpose of homotopical investigations, the following direct consequence of the *local product* structure of fibre bundles is of fundamental importance.

Homotopy lifting property. *Let $B \xrightarrow{p} X$ be a fibre bundle and K be a finite polyhedron. Let $f_0: K \to B$ be a map and $\bar{f}: K\times I \to X$ be a homotopy of $p\cdot f_0 = \bar{f}_0$. Then there exists a homotopy $f: K\times I \to B$ covering $\bar{f}$* (i.e., $pf=\bar{f}$), *and f is stationary with $\bar{f}$.*

The proof of the above property is simple and straightforward by using the local product structure of $p: B\to X$. However, it is so fundamental that it led J.P. Serre to axiomatize it in defining fibre spaces [S 3].

Induced bundle. Let $B \xrightarrow{p} X$ be a fibre bundle and $f: X' \to X$ be a map. Let B' be those points (b, x') of $B\times X'$ with $p(b)=f(x')$ and $\tilde{f}, p'$ be the restriction of projections to B', i.e.,

$$\begin{array}{ccc} B\times X' & & \\ \cup\!\!\!\mid & & \\ B' & \xrightarrow{\tilde{f}} & B \\ \Big\downarrow p' & & \Big\downarrow p \\ X' & \xrightarrow{f} & X\,. \end{array}$$
$$B\times X' \to B, \qquad B\times X' \to X'$$

Then, it is easy to check that $B' \xrightarrow{p'} X'$ is a fibre bundle and $\tilde{f}$ is a bundle map. $B' \xrightarrow{p'} X'$ is called the *induced bundle* of $B \xrightarrow{p} X$ with respect to f.

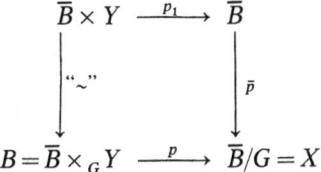

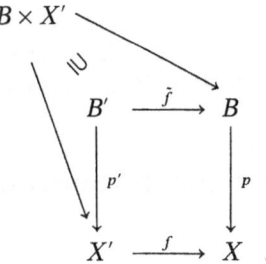

Classification Theorem. *Let $E_G^n \xrightarrow{p} B_G^n$ be a principal G-bundle with $\pi_i(E_G^n)=0$ for $0 \leqslant i \leqslant n$, and K be a finite polyhedron of $\dim \leqslant n$. Then the operation of assigning to each map $f: K \to B_G^n$ its induced bundle sets up a bijection between homotopy classes of maps of K into B_G^n and equivalence classes of principal G-bundles over K.*

We refer to § 19 of [S 11] and [M 2] for a proof of the above theorem and the general existence of such classifying (universal) bundle.

(E) *Reduction and Extension of Structural Group*

Let $H \subseteq G$ be a closed subgroup of G and $B \xrightarrow{p} X$ and $B' \xrightarrow{p'} X$ be respectively H-bundle and G-bundle over X (i.e., free H-space and G-space with X as their common orbit space). If there exists an equivariant map $r: B \to B'$ which commutes with projections, i.e.,

$$\begin{array}{ccc} B & \xrightarrow{r} & B' \\ & {\scriptstyle p}\searrow \quad \swarrow{\scriptstyle p'} & \\ & X & \end{array}$$

commutes and $r(b \cdot h) = r(b) \cdot h$, then $B \xrightarrow{p} X$ is called a reduction of $B' \xrightarrow{p'} X$ and $B' \xrightarrow{p'} X$ is called an extension of $B \xrightarrow{p} X$.

Observe that $B' = B' \times_G G$ and hence $B'/H = B' \times_G G/H$. Therefore, a reduction $r: B \to B'$ induces the following maps

$$\begin{array}{ccc} B/H & \xrightarrow{\bar{r}} & B'/H = B' \times_G G/H \\ & {\scriptstyle \bar{p}}\searrow \cong \quad \swarrow{\scriptstyle \bar{p}'} & \\ & X & \end{array}$$

which gives a cross-section $\bar{r} \cdot \bar{p}^{-1}: X \to B' \times_G G/H$. Conversely, suppose $s: X \to B' \times_G G/H = B'/H$ is a given cross-section. Then the inverse image of $s(X)$ of the projection $p_0: B' \to B'/H$, i.e., $B = p_0^{-1}(s(X))$, is clearly a free H-space with $s(X) = X$ as its orbit space and the inclusion $B = p_0^{-1}(s(X)) \subseteq B'$ is equivariant. Hence,

Proposition. *For a given closed subgroup $H \subseteq G$, there is a $1-1$ correspondence between the reductions of a G-bundle $B' \to X$ to an H-bundle and the cross-sections of the associated G/H-bundle $B'/H = B' \times_G G/H \to X$.*

§ 3. The Existence of Slice and its Consequences on General G-Space

From now on, we shall always assume that G is a compact Lie group. The following simple, useful fact was first noticed by Koszul [K 3] and then further generalized and exploited by Montgomery-Yang, Mostow and Palais. [M 4, M 9, P 1]

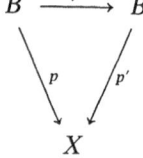

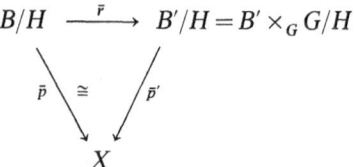

(A) *Differentiable Slice*

Theorem (I.5). *Let M be a differentiable G-space, $H=G_x$ be the isotropy subgroup at a point $x\in M$, and φ_x be the local representation of H on normal vectors (w.r.t. a chosen invariant metric) of the orbit $G(x)\cong G/H$ at x. Then, the equivariant normal bundle $\nu(G(x))$ of $G(x)$ is isomorphic to*

$$\alpha(\varphi_x)\colon \mathbb{R}^k \to G\times_H \mathbb{R}^k \to G/H$$

where H acts on G as right translations and on $\mathbb{R}^k$ via φ_x. Furthermore, for a sufficient small $\varepsilon>0$, the exponential map (w.r.t. a chosen invariant metric) is an equivariant diffeomorphism of the associated ε-disc bundle of $\alpha(\varphi_x)\colon G\times_H D^k_\varepsilon \to G/H$ onto an invariant tubular neighborhood of $G(x)$.

Proof. Let $\mathbb{R}^k_x$ be the set of normal vectors to $G(x)$ at x (w.r.t. a chosen invariant metric, which always exists by means of averaging). Then the induced G-action carries $\mathbb{R}^k_x$ to $\mathbb{R}^k_{g\cdot x}$. It is easy to check that $g_1\cdot v_1=g_2\cdot v_2$ if and only if (g_1,v_1) and $(g_2,v_2)\in G\times\mathbb{R}^k_x$ are equivalent in the usual sense, i.e.,

$$\begin{array}{ccccc} G\times\mathbb{R}^k_x & \subseteq & G\times T(M) & \longrightarrow & T(M)\\ \downarrow & & \searrow & & \cup \\ G\times_H\mathbb{R}^k_x & & \xrightarrow{\quad\cong\quad} & & \nu(G(x)) \end{array}$$

Hence, the above theorem follows from a straightforward verification. ☐

(B) *Gleason Lemma and the Existence of Topological Slice*

Lemma (3.1). *If H is a closed subgroup of G (a compact Lie group), then there exists a finite dimensional linear G-action with a vector v such that $H=G_v$ is the isotropy subgroup of v.*

Proof. It follows directly from Peter-Weyl theorem and the descending chain condition for closed subgroups of a compact Lie group. We refer to the paper of Mostow [M 9] and Palais [P 1] for a detail proof. ☐

Gleason Lemma (Generalized Form). *If K is a compact invariant subspace of a G-space X and if $f\colon K\to V$ is an equivariant map of K into a linear G-space, then f admits an equivariant extension $\tilde{f}\colon X\to V$.*

Proof. By Tietze's extension theorem, there is a continuous extension $f^*\colon X\to V$. Let $\tilde{f}(x)=\int_G g^{-1} f^*(gx)\,dg$. It is easy to check that $\tilde{f}$ is an equivariant extension of f.

Theorem (I.5′) (Existence of a slice). *Let X be a topological G-space and $H=G_x$. Then there exists a subset S of X such that* (i) *S is invariant under H,* (ii) *$G(S)=G\times_H S$ and is an invariant neighborhood of $G(x)$* in *X. Such a subset is called a slice at x.*

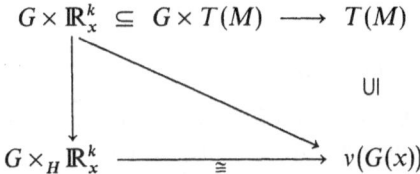

Proof. In the special case that X is a differentiable G-space, a small normal disc $D_x^k(\varepsilon)$ is clearly such a slice at x. In the general topological case, we first equivariantly embed the orbit $G(x)\cong G/H$ into a suitable linear G-space $V, f: G(x) \cong G/H \xrightarrow{\cong} G(v)\subseteq V$ (by lemma (3.1)). Then extend f to an equivariant map $\tilde{f}: X\to V$. Let S' be a slice at v (which exists because V is linear) and put $S=\tilde{f}^{-1}(S')$. Then it is easy to check that S is such a slice. □

The above slice theorem has the following two consequences of basic importance:

(C) *Equivariant Embedding*

Theorem (I.6) (Mostow-Palais). *If G is a compact Lie group and X is a separable, metrizable G-space of finite dimension and with finite orbit types, then X admits an equivariant imbedding into a linear G-space.*

Since we actually do not need the above theorem in any essential way, we refer to [M 9] or Ch. VIII of [B 10] for a proof.

(D) *Principal Orbit Type Theorem*

For a given G-space X, the set of all isotropy subgroups $\mathcal{O}(X)=\{G_x; x\in X\}$ clearly divides into conjugacy classes which correspond to the types of orbits in X. Observe that a homogeneous space G/K can map equivariantly onto G/H if and only if K is conjugate to a subgroup of H, i.e., $gKg^{-1}\subseteq H$ for suitable $g\in G$. Hence, it is natural to introduce the following partial ordering relation among the set of orbit types:

$$\mathcal{O}(X)=\{G_x : x\in X\}=\bigcup (H_i) \text{ (conjugacy classes)},$$
$$(H_i)\geqslant(H_j)\Leftrightarrow H_i \text{ is conjugate to a subgroup of } H_j.$$

Theorem (I.7) (Montgomery-Samelson-Yang). *Let M be a connected manifold with a given differentiable G-action. Then, among the set of orbit types $\mathcal{O}(M)=\bigcup(H_i)$, there is a unique maximal orbit type (H_1) such that*

(i) $(H_1)\geqslant(G_x)$ *for all* $x\in M$,

(ii) *the union of all orbits of* (H_1)-*type* $= M_{(H_1)}=\{x; G_x\in(H_1)\}$ *is open dense in M and the codimension of* $(M-M_{(H_1)})$ *is at least* 1.

(iii) $F(H_1, M)$ *intersects every orbit,*

(iv) *the orbit space* $M_{(H_1)}/G$ *is connected.*

Proof. Let $M'=M/G$ be the orbit space and $\mathcal{O}(M)$ be the set of orbit types (with discrete topology and partial ordering). Let us consider the orbit type function $\Omega: M'\to\mathcal{O}(M)$. It follows from the differentiable slice theorem that $x'=G(x)$ is a local maximal if and only if φ_x is a *trivial* representation (i.e., G_x acts trivially on the slice $S_x=D_x^k$, and hence Ω is *locally constant* in a neighborhood of those *local maximal* points x'. On the other hand, suppose $y'=G(y)$ is *not* a local maximal, then there are the following two cases:

(1) codim $(F(G_y, S_y)) \geq 2$, then $S_y - F(G_y, S_y)$ is still connected and hence if we remove those orbits of the same type as y' from the neighborhood of y', S_y/G_y, the remaining set $[S_y - F(G_y, S_y)]/G_y$ is still *connected.*

(2) codim $F(G_y, S_y) = 1$, then $\varphi_y: G_y \to 0(1) \cong \mathbb{Z}_2$ and acts on S_y as a reflection with respect to the hyperplane $F(G_y, S_y)$. Hence the orbit space S_y/G_y is a half plane with the image of $F(G_y, S_y)$ as boundary. Therefore $[S_y - F(G_y, S_y]/G_y$ is again *connected.*

Since M is assumed to be connected, $M' = M/G$ must be also *connected.* The above fact shows that the *removing* of all those points y' which are *not* local maximals did not even separate the space M' locally. Hence, the subspace M_1' of all local maximal points x' of M' is still *connected.* But Ω is *locally constant* on M_1', Ω must be in fact a *constant* on M_1'. That is, there is a *unique* (local) maximal orbit type (H_1) and $M_1' = M_{(H_1)}/G$ is connected, open dense in M'. The assertions (ii), (iii) also follow immediately. □

Remark. (i) The above theorem also holds without modification for topological G-action on cohomology manifold over $\mathbb{Z}$. It also holds for connected orbit types, (i.e., the types of connected isotropy subgroups G_x^0) for topological G-action on cohomology manifold over $\mathbb{Q}$. Such generalizations will be proved when we are ready to show the same fact about codim $F(G_y, S_y)$ by cohomology method.

(ii) The unique maximal orbit type is called the *principal orbit type*, and its corresponding isotropy subgroups are called *principal isotropy subgroups.*

§ 4. General Theory of Compact Connected Lie Groups

(A) *Adjoint Action and Adjoint Representation*

A *one-parameter subgroup* of a Lie group G is a (differentiable) homomorphism, ξ, of the simplest Lie group $\mathbb{R}^1$ into G. Clearly, the *right* translations of $\{\xi(t), t \in \mathbb{R}^1\}$ exhibit a *left-invariant* (due to the associativity of G) $\mathbb{R}^1$-action on G. Hence, the velocity vector at every point forms a *left-invariant* vector field X on G. Conversely, if we integrate a left invariant vector field X, then it is easy to show (by uniqueness and left invariance of integral curves) that the integral curve passing through identity is a one-parameter subgroup. Hence, there are the following bijections

$$\begin{array}{ccc} \{\text{one-parameter subgroups}\} & \leftrightarrow & \{\text{left invariant } \mathbb{R}^1\text{-actions}\} \\ \updownarrow & & \updownarrow \\ \{\text{tangent vectors at identity}\} & \leftrightarrow & \{\text{left invariant vector fields}\}. \end{array}$$

We identify them via the above bijections and call it the *Lie algebra*, $\mathfrak{g}$, of G, which is a vector space with a bilinear bracket product $[X, Y]$ (of vector fields) satisfying the following Jacobi identity:

$$[[X, Y], Z] + [[Y, Z], X] + [[Z, X], Y] \equiv 0.$$

Furthermore, it is convenient and useful to organize all one-parameter subgroups into a (universal) map:

$$\mathrm{Exp}:\mathfrak{g}\to G$$

such that $\mathrm{Exp}(tX):\mathbb{R}^1\to\mathfrak{g}\to G$ gives the one-parameter subgroup with X as its velocity vector at the identity e.

Remark. (i) Recall that the geometric meaning of bracket product $[X,Y]$ of two vector fields X,Y is the following: If one "drifts" successively in the flow $X,Y,(-X)$ and then $(-Y)$ each for a time t, one ends up travelling approximately $t^2\cdot[X,Y]$. Hence, in our case of left-invariant vector fields, we have $[X,Y]_e=$ the velocity vector (at e) of the curve

$$\gamma(t)=\mathrm{Exp}\sqrt{t}X\cdot\mathrm{Exp}\sqrt{t}Y\cdot\mathrm{Exp}(-\sqrt{t}X)\cdot\mathrm{Exp}(-\sqrt{t}Y).$$

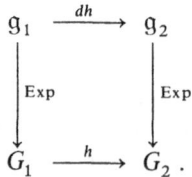

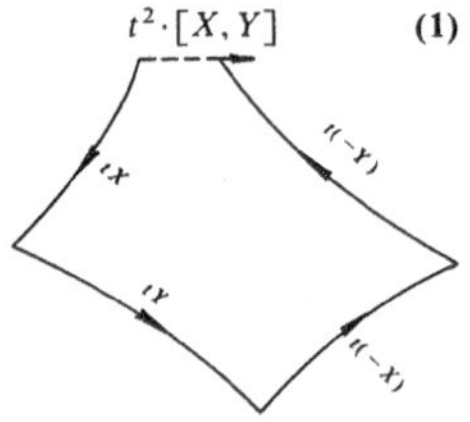

(ii) If $h:G_1\to G_2$ is a homomorphism of Lie groups, then the differential of h at e clearly induces a homomorphism of their Lie algebra $dh_e:\mathfrak{g}_1\to\mathfrak{g}_2$,

$$\begin{array}{ccc}\mathfrak{g}_1 & \xrightarrow{dh} & \mathfrak{g}_2\\ \downarrow{\scriptstyle \mathrm{Exp}} & & \downarrow{\scriptstyle \mathrm{Exp}}\\ G_1 & \xrightarrow{h} & G_2\,.\end{array}$$

Definition. The conjugations represent G as a differentiable transformation on itself via inner automorphisms, i.e., $G\times G\to G$, $(g_1,g_2)\mapsto g_1g_2g_1^{-1}$. We shall call it the *adjoint* G-action on G itself. The above adjoint action on G induces another adjoint action on its Lie algebra $\mathfrak{g}$, which is a linear representation of G into $\mathrm{GL}(\mathfrak{g})$, $\mathrm{Ad}:G\to\mathrm{GL}(\mathfrak{g})$ (general linear group of $\mathfrak{g}$), called the adjoint representation of G. Furthermore, the differential of Ad at e, $\mathrm{ad}=\mathrm{Ad}_e:\mathfrak{g}\to\mathfrak{g}l(\mathfrak{g})$ is called the adjoint representation of $\mathfrak{g}$. In terms of exponential maps, the above two adjoint representations of G and $\mathfrak{g}$ are respectively characterized by the following identities

(i) $\mathrm{Exp}\,t\cdot[\mathrm{Ad}(\mathrm{Exp}\,sX)\cdot Y]=\mathrm{Exp}\,sX\cdot\mathrm{Exp}\,tY\cdot\mathrm{Exp}(-sX)$,
(ii) $\mathrm{Ad}(\mathrm{Exp}\,tX)\cdot Y=\mathrm{Exp}(t\cdot\mathrm{ad}(X))\cdot Y$.

Proposition (4.1). $\mathrm{ad}(X)\cdot Y=[X,Y]$ *for all* $X,Y\in\mathfrak{g}$.

Proof. It follows from the geometric interpretation of the bracket product that

$$\mathrm{Exp}\,sX\;\mathrm{Exp}\,tY\cdot\mathrm{Exp}(-sX)\cdot\mathrm{Exp}(-tY)=\mathrm{Exp}\,st\cdot[X,Y]+0(st).$$

On the other hand, by the above definition, we have

$$
\begin{aligned}
&\mathrm{Exp}[(\mathrm{Exp}\, s\, \mathrm{ad}(X) - I)\cdot t\, Y] = \mathrm{Exp}(st\cdot \mathrm{ad}(X)\cdot Y) + 0(st)\\
&\qquad\qquad \Big\|\\
&\mathrm{Exp}(t\, \mathrm{Exp}\, s\, \mathrm{ad}(X)\cdot Y)\cdot \mathrm{Exp}(-t\, Y) + 0(st)\\
&\qquad\qquad \Big\|\ \text{by (ii)} \qquad\qquad \text{by (i)}\\
&\mathrm{Exp}\, t[\mathrm{Ad}(\mathrm{Exp}\, sX)\cdot Y]\,\mathrm{Exp}(-t\, Y) = \mathrm{Exp}\, sX\cdot \mathrm{Exp}\, t\, Y\cdot \mathrm{Exp}(-sX)\,\mathrm{Exp}(-t\, Y)\\
&\qquad\qquad\qquad\qquad\qquad\qquad = \mathrm{Exp}\, st\cdot [X, Y] + 0(st).
\end{aligned}
$$

Hence, if we take the limit $s, t \to 0$, we get $\mathrm{ad}(X)\cdot Y = [X, Y]$. □

(B) *Cartan's Theorem of Maximal Tori and its Consequences*

Theorem (I.8) (É. Cartan). *Let G be a compact connected Lie group and T be a given maximal torus. Then*

(i) *all maximal tori of G are mutually conjugate and they are exactly those connected principal isotropy subgroups of the adjoint action on G.*

(ii) *T intersects with every conjugacy class (i.e., orbit of adjoint action) or equivalently $G = \bigcup \{g T g^{-1}; g \in G\}$.*

Proof. The above theorem is a direct consequence of Theorem (I.7) and the fact that maximal tori are exactly those connected principal isotropy subgroups of the adjoint action on G. Since principal orbits are everywhere dense and the adjoint action *on* $\mathfrak{g}$ is precisely the local linear action around the identity $e \in G$, it is equivalent but technically simpler to prove that the maximal tori are the connected principal isotropy subgroups of the *adjoint action* on $\mathfrak{g}$. The following lemma provides an effective way of computing the principal isotropy subgroups of a linear action.

Lemma (4.1). *Let ψ be a linear G-action and φ_x be the slice representation of G_x at x* (cf. § 3—A). *Then*

$$\varphi_x + (\mathrm{Ad}_G | G_x - \mathrm{Ad}_{G_x}) = \psi | G_x .$$

Proof. It is clear that the local representation of G_x at origin is $\psi | G_x$, while that of x is φ_x (: normal part) + $(\mathrm{Ad}_G | G_x - \mathrm{Ad}_{G_x})$ (: tangent part). But the fixed point set $F(G_x, V)$ is a linear subspace and hence obviously connected. Therefore the local representations of G_x at origin and at x must be equal. □

Let X be an arbitrary element of $\mathfrak{g}$, G_X^0 be the connected isotropy subgroup of X and $\mathfrak{g}_X$ be the Lie algebra of G_X. It is easy to show the existence of at least a maximal torus $T \supseteq \mathrm{Exp}\, tX$, hence G_X contains at least one maximal torus for any $X \in \mathfrak{g}$. On the other hand, it follows from the above lemma that G_X^0 is a connected principal isotropy subgroup if and only if

$$\varphi_X = \mathrm{Ad}_G | G_X^0 - (\mathrm{Ad}_G | G_X^0 - \mathrm{Ad}_{G_X^0}) = \mathrm{Ad}_{G_X^0}$$

is a trivial representation which (in view of the fact G_X contains at least a maximal torus), is equivalent to saying that G_X^0 *itself is a maximal torus.* □

Corollary (I.8.1). $\mathrm{Exp}: \mathfrak{g} \to G$ *is onto for a compact connected Lie group.*

Proof. Let $g \in G$ be an arbitrary element then there exists $g_1 \in G$ such that $g_1 g g_1^{-1} \in T$ or $g \in g_1^{-1} T g_1$ and hence, the above corollary follows from the obvious fact that Exponential map is onto for a torus.

Corollary (I.8.2). G_X *(w.r.t. the adjoint action on* $\mathfrak{g}$*) is connected for every* $X \in \mathfrak{g}$.

Proof. Observe that G_X = centralizor of the torus subgroup $T(X)$ generated by $\{\mathrm{Exp}\, tX\}$. We claim that G_X = the union of all maximal tori containing $T(X)$ and hence connected. Let $g \in G_X$ be an arbitrary element of G_X. Then $g, T(X)$ generates a compact abelian subgroup which is, in general, a cyclic group product with a torus, i.e., $\overline{\langle g, T(X)\rangle} = \mathbb{Z}_h \times T_1$. It is easy to show that $\mathbb{Z}_h \times T_1 = \overline{\langle g_1 \rangle}$ is a *topological cyclic group*. Hence, by the above theorem, there exists a maximal torus $T \supseteq \overline{\langle g_1 \rangle} = \overline{\langle g, T(X)\rangle}$, i.e., $g \in T \supseteq T(X)$. □

Corollary (I.8.3). *Let* $T \subseteq G$ *be a maximal torus of a compact connected Lie group* G*. Then two representations* φ, ψ *of* G *are equivalent if and only if their restrictions to* T*,* $\varphi|T$ *and* $\psi|T$ *are equivalent.*

Proof. $\varphi \sim \psi \Leftrightarrow \chi_\varphi = \chi_\psi \Leftrightarrow \chi_\varphi|T = \chi_{\varphi|T} = \chi_{\psi|T} = \chi_\psi|T \Leftrightarrow \varphi|T \sim \psi|T$
(for character functions take constant value on each conjugacy class and T intersects every conjugacy class). □

Definition. The dimension of maximal tori of a compact Lie group G is called the *rank* of G.

(C) *Root System and Weight System*

Definition. For a given complex representation ψ of a compact connected Lie group G and a fixed maximal torus T, it follows from the above corollary and Schur lemma that $\psi|T$ is a *complete invariant* and

$$\psi|T = \psi_1 \oplus \psi_2 \oplus \cdots \oplus \psi_n$$

splits uniquely into the sum of one-dimensional representations. Hence, the collection of integral linear functionals $w_i \in H^1(T, \mathbb{Z})$ corresponding to ψ_i forms a complete invariant of ψ, called the weight system of ψ, and denoted by $\Omega(\psi)$.

Definition. The (non-zero) *weight system* of the complexification of the adjoint representation of a compact connected Lie group G is called the root system of G, and is denoted by $\Delta(G)$, i.e.,

$$\Delta(G) = \Omega(\mathrm{Ad}_G \otimes \mathbb{C}) - \{\text{those zero weight vectors}\}.$$

Remark. The exclusion of zero weights in $\Delta(G)$ is purely for notational convenience.

Chapter II. Structural and Classification Theory of Compact Lie Groups and Their Representations

This chapter consists of a concise exposition of the structure and classification theories of compact Lie groups and their representations from the geometric viewpoint of transformation groups. An explicit and neat understanding of the orbit structure of the adjoint action of a compact Lie group G plays a central role in the classification theory developed by É. Cartan and H. Weyl. This more geometric approach is actually more natural and straightforward than the usual Lie-algebra-theoretical approach. Furthermore, such an approach will also provide us with valuable examples and insight for later investigation of topological transformation groups.

§ 1. Orbit Structure of the Adjoint Action

(A) *The simplest fundamental examples:* SU(2) *or* SO(3)

The special unitary group of rank 2, SU(2) consists of those 2×2 matrices $\sigma=\begin{pmatrix} a & -\bar{b} \\ b & \bar{a} \end{pmatrix}$ with $\det(\sigma)=|a|^2+|b|^2=1$. It can also be identified with the group of unit quaternions, $\mathrm{Sp}(1)=\{q=a+j\cdot b, |q|^2=|a|^2+|b|^2=1\}$, which is geometrically the unit sphere $S^3(1)\subseteq\mathbb{R}^4$. The center of Sp(1) consists of $\{\pm 1\}$ and the adjoint representation of Sp(1) maps $\mathrm{Sp}(1)/\{\pm 1\}$ *isomorphically* onto SO(3), the rotation group of euclidean 3-space. Geometrically, the adjoint action on $\mathrm{Sp}(1)=S^3(1)$ is simply the SO(3)-rotation with the real axis as the fixed axis. Since the total volume of the unit 3-sphere is $2\pi^2$, the *normalized Haar measure* on Sp(1) is the usual measure on $S^3(1)$ modified by a constant $1/2\pi^2$.

Now let us apply the above understanding of the geometric structure of $\mathrm{Sp}(1)=\mathrm{SU}(2)$ to the classification of irreducible complex representations of Sp(1).

(i) Let $\psi_1:\mathrm{Sp}(1)\xrightarrow{\cong}\mathrm{SU}(2)$ be the usual complex linear action of SU(2) on $\mathbb{C}^2=\{(z_1,z_2)\}$. Then the above linear substitutions induce a complex linear action, ψ_k, of $\mathrm{Sp}(1)\cong\mathrm{SU}(2)$ on the space of degree k homogeneous polynomials $P_k\subseteq\mathbb{C}[z_1,z_2]$ for each $k\geqslant 0$.

(ii) Let $T^1=\{a=e^{2\pi i\theta}\}$ = unit complexes $\subseteq$ Sp(1) (unit quaternions). Then T^1 is a maximal torus of Sp(1) and $\psi_1(a)=\begin{pmatrix} a & 0 \\ 0 & a^{-1} \end{pmatrix}$, i.e.,

$$\psi_1(a)\cdot z_1=a\cdot z_1, \quad \psi_1(a)\cdot z_2=a^{-1}\cdot z_2.$$

Therefore, for the basis $\{z_1^k,\dots,z_1^{k-l}z_2^l,\dots,z_2^k\}$ of P_k,

$$\psi_k(a)\cdot(z_1^{k-l}z_2^l)=(a\,z_1)^{k-l}\cdot(a^{-1}z_2)^l=a^{k-2l}(z_1^{k-l}\cdot z_2^l)$$

i.e., $\psi_k(a)=\operatorname{diag}(a^k,a^{k-2},\dots,a^{(k-2l)},\dots,a^{-k})$ and

$$\chi_k(a)=\chi_{\psi_k}(a)=\operatorname{tr}\psi_k(a)=a^k+a^{k-2}+\dots+a^{-k}=\frac{(a^{k+1}-a^{-(k+1)})}{(a-a^{-1})}.$$

Theorem (II.1). $\{\psi_k,k\geqslant 0\}$ *constitutes a complete collection of (non-equivalent) irreducible representations of* $\mathrm{Sp}(1)=\mathrm{SU}(2)$.

Proof. Let $d\sigma$ be the usual measure on $S^3(1)$ and f be a function of $G=\mathrm{Sp}(1)$ constant on every conjugacy class (i.e., a *central function*). Then, we have the following integration formula

$$\begin{aligned}\int_G f(g)\,dg&=\frac{1}{2\pi^2}\int_{S^3}f(g)\,d\sigma=\frac{1}{2\pi^2}\int_0^{1/2}f(e^{2\pi i\theta})\cdot 4\pi\sin^2(2\pi\theta)\cdot 2\pi\,d\theta\\&=\tfrac{1}{2}\int_0^1 f(e^{2\pi i\theta})\cdot|e^{2\pi i\theta}-e^{-2\pi i\theta}|^2\cdot d\theta=\tfrac{1}{2}\int_{T^1}f(t)\cdot|Q(t)|^2\,dt\end{aligned}$$

where "dg" and "dt" are the normalized Haar measure of $G=\mathrm{Sp}(1)$ and T^1 respectively and $Q(t)=(t-\bar t)$ for $t=e^{2\pi i\theta}\in T^1$.

(i) Applying the above formula to $f(g)=\chi_k(g)\cdot\overline{\chi_k(g)}$, one gets

$$\begin{aligned}\int_G\chi_k\cdot\bar\chi_k\,dg&=\tfrac{1}{2}\int_{T^1}\chi_k(t)\cdot\overline{\chi_k(t)}\cdot Q(t)\cdot\overline{Q(t)}\,dt\\&=\tfrac{1}{2}\int_{T^1}|\chi_k(t)\cdot Q(t)|^2\,dt\\&=\tfrac{1}{2}\int_{T^1}|t^{k+1}-t^{-(k+1)}|^2\,dt=1\,.\end{aligned}$$

Hence, it follows from Theorem (I.3′) that ψ_k are irreducible for all $k\geqslant 0$.

(ii) Notice that $\dim\psi_k=(k+1)$ are different for different k, it is obvious that $\{\psi_k,k\geqslant 0\}$ form a non-equivalent family. On the other hand, it is a well-known fact in Fourier series that

$$\{(e^{2\pi i(k+1)\theta}-e^{-2\pi i(k+1)\theta})=\chi_k(e^{2\pi i\theta}-e^{-2\pi i\theta}),\ k\geqslant 0\}$$

forms a basis of the sub-space of *odd functions* in $L^2(T^1)$. Hence, $\{\psi_k,k\geqslant 0\}$ is already a complete family in the sense that any irreducible complex representation, ψ, of $\mathrm{Sp}(1)$ must be equivalent to one of ψ_k. For otherwise, it follows from Theorem (I.3′) that

$$\begin{aligned}\langle\chi_\psi,\chi_k\rangle_{L^2(G)}&=\int_G\chi_\psi\cdot\bar\chi_k\,dg\\&=\tfrac{1}{2}\int_T\chi_\psi(t)Q(t)\cdot\overline{\chi_k(t)\cdot Q(t)}\,dt=0\quad\text{for all }k\,.\end{aligned}$$

Since $\chi_\psi(t)\cdot Q(t)$ is a *continuous odd-function*, the above orthogonal relations for the whole basis $\{\chi_k(t)\cdot Q(t)\}$ imply that $\chi_\psi(t)\cdot Q(t)\equiv 0$, which is obviously a contradiction, for $\chi_\psi(e)=\dim\psi\neq 0$. □

Remark. (i) The weight system of ψ_k form a string leading from the highest weight vector $k\theta$ to $-k\theta$, i.e.,

$$\Omega(\psi_k)=\{k\theta,(k-2)\theta,\ldots,(k-2l)\theta,\ldots,-k\theta\}.$$

(ii) ψ_k factor through SO(3), i.e., $\mathrm{Ker}(\psi_k)=\{\pm 1\}$, if and only if k is even. And they forms a complete family of irreducible representations of SO(3).

(iii) In either case of Sp(1), or SO(3), the weight system of any representation consists of at least one non-negative weight which is *less* than the positive root, i.e., $0\leqslant\omega<\alpha$.

Corollary (II.1.1). *Every non-commutative compact connected Lie group of rank one is either isomorphic to* Sp(1) *or to* SO(3).

Proof. Observe that the Lie algebras of Sp(1) and SO(3) are *isomorphic* and can be expressed in terms of basis $\mathfrak{g}_1=(H,X,Y)$ with $[H,X]=Y$, $[H,Y]=-X$, and $[X,Y]=H$. Now, let G be a *non-commutative compact connected* Lie group of *rank one* and $\mathfrak{g}$ be its Lie algebra. Let $T^1\subseteq G$ be an arbitrary, fixed maximal torus of G, and $\mathrm{Ad}|T^1$ be the restriction of the adjoint representation to T^1. Then

$$\mathrm{Ad}|T^1=1+\textstyle\sum_{j=1}^s\alpha_j \quad\text{and}\quad \mathfrak{g}=\mathbb{R}^1+\textstyle\sum_{j=1}^s\mathbb{R}^2_{\alpha_j}$$

where $\mathbb{R}^1$ is the Lie algebra of T^1 and $\alpha_j\colon T^1\to\mathrm{SO}(2)$ is a homomorphism with winding number $n_j>0$; $n_1\leqslant\cdots\leqslant n_s$. Let $\mathfrak{g}_1'=\mathbb{R}^1+\mathbb{R}^2_{\alpha_1}$ and H' the integral base vector of $\mathbb{R}^1$, i.e., $\mathbb{Z}\cdot H'=\mathrm{Ker}(\mathbb{R}^1\xrightarrow{\exp}T^1)$, and X',Y' be an orthogonal basis of $\mathbb{R}^2_{\alpha_1}$. Then, by definition,

$$\alpha_1(\mathrm{Exp}\,t\cdot H')=\begin{pmatrix}\cos 2\pi n_1 t & -\sin 2\pi n_1 t\\ \sin 2\pi n_1 t & \cos 2\pi n_1 t\end{pmatrix}.$$

Hence, by Proposition (4.1) of § I.4—A, we have

$$[H',X']=n_1 Y' \quad\text{and}\quad [H',Y']=-n_1 X'.$$

The fact that G is of rank one implies that $[A,B]=0$ if and only if A,B are *linearly dependent* in $\mathfrak{g}$. Hence $[X',Y']\neq 0$, and

$$\begin{aligned}[H',[X',Y']] &= -[X',[Y',H']]-[Y',[H',X']]\\ &= -[X',n_1X']-[Y',n_1Y']=0\end{aligned}$$

imply that $[X',Y']=\lambda\cdot H'$ for a suitable $\lambda\neq 0$. Then, it is easy to modify the basis $\{H',X',Y'\}$ by suitable constant to obtain a basis $\{H_1,X_1,Y_1\}$ of $\mathfrak{g}_1'$ so that $[H_1,X_1]=Y_1$, $[H_1,Y_1]=-X_1$, $[X_1,Y_1]=H_1$. Hence $\mathfrak{g}_1'\cong\mathfrak{g}_1$ as Lie algebra and consequently, the subgroup G_1' with $\mathfrak{g}_1'$ as its Lie algebra is isomorphic to either Sp(1) or SO(3). We claim that $G_1'=G$. For otherwise, it follows by applying the above theorem to the isotropy representation of G_1' on tangent vectors of G/G_1' that its weight system $\Omega=\{\alpha_2,\dots,\alpha_s\}$ must contain at least one $0\leqslant\alpha_j<\alpha_1$ (cf. Remark (iii)) which is a contradiction to the choice that $\alpha_1\leqslant\alpha_j$ for all $2\leqslant j\leqslant s$. □

(B) *An Elementary Structural Theorem for Groups Generated by Reflections*

Let M be a connected differentiable manifold. A diffeomorphism r of M onto itself is called a reflection if $r^2=$ identity, the fixed point set of r, $F(r)$ is of codimension one and $(M-F(r))$ consists of two connected components interchanged by r.

Theorem (II.2). *Let Γ be a finite group generated by reflections and $\mathcal{R}=\{r_j\}$ be the set of all reflections in Γ. Then Γ acts simply transitively on the connected components of $(M-\bigcup\{F(r_j); r_j\in\mathcal{R}\})$ (called chambers). The closure of an arbitrary chamber $\bar{C}_0$ is a fundamental domain of Γ in the sense that every point $x\in M$ is conjugate to exactly one point $x_0=\gamma\cdot x\in C_0$.*

Proof. Observe that $\gamma\cdot F(r)=F(\gamma\cdot r\cdot\gamma^{-1})$ and hence $\bigcup\{F(r_j), r_j\in\mathcal{R}\}$ is invariant under Γ, and consequently $(M-\bigcup\{F(r_j), r_j\in\mathcal{R}\})$ is also invariant. Suppose C_i and C_j are two chambers with $(\bar{C}_i\cap F(r)\cap\bar{C}_j)$ of *codimension one.* Then it is obvious that $r(C_i)=C_j$. Since the *deletion* of all *intersections* of different hyperplanes $\{F(r_i)\cap F(r_j), r_i\neq r_j\in\mathcal{R}\}$, (which is of codimension $\leqslant 2$), fails to disconnect M, it is not difficult to show that Γ acts transivitely on the set of all chambers, i.e., $\Gamma\cdot C_0=(M-\bigcup\{F(r_j), r_j\in\mathcal{R}\})$. Hence $\Gamma\cdot\bar{C}_0=\overline{(M-\bigcup\{F(r_j)\})}=M$. We refer to [H8] for a detail proof that $\bar{C}_0$ intersects every Γ-orbit *exactly once*, i.e., $\bar{C}_0$ is a fundamental domain. □

Corollary (II.2.1). *Let $\mathcal{R}_0$ be the set of those reflections in Γ with their hyperplanes $F(r)$ have codimension one intersection with $\bar{C}_0$. Then Γ is already generated by those reflections $\{r\in\mathcal{R}_0\}$, called a simple system of generators of Γ.*

(C) *Weyl Group and a Fundamental Reduction*

Definition. Since all maximal tori of a compact connected Lie group G are conjugate, we may arbitrarily choose a maximal torus $T\subseteq G$ and define the Weyl group of G, $W(G)=N(T)/T$, where $N(T)$ is the normalizor of T in G. Clearly, the conjugations induce an action of $W(G)$ on T and consequently also on its Lie algebra $\mathfrak{h}$, the Cartan subalgebra. The importance of the Weyl group lies in the following fundamental reduction.

Theorem (II.3). *Let G be a compact connected Lie group, T a maximal torus of G, $\Delta(G)$ the root system of G, and W the Weyl group of G. Then*

(i) *the multiplicity of each root $\alpha\in\Delta(G)$ is one and $k\cdot\alpha\in\Delta(G)$ iff $k=\pm 1$,*

(ii) *the action of W on $\mathfrak{h}$ (resp. on T) is a group generated by reflections $\{r_\alpha, \alpha\in\Delta(G)\}$, where r_α is the reflection w.r.t. the hyperplane $H_\alpha=\mathrm{Ker}(\mathfrak{h}\xrightarrow{\alpha}\mathbb{R}^1)$* (resp. $T_\alpha=\mathrm{Ker}(T\xrightarrow{\alpha}U(1))$),

(iii) $W_x = W(G_x)$ *and the following inclusions induce bijections on their respective orbit spaces, i.e.,*

$$\begin{array}{ccc} \mathfrak{h} & \xrightarrow{\subseteq} & \mathfrak{g} \\ \downarrow & & \downarrow \\ \mathfrak{h}/W & \xrightarrow{\cong} & \mathfrak{g}/\mathrm{Ad} \end{array} \qquad \begin{array}{ccc} T & \xrightarrow{\subseteq} & G \\ \downarrow & & \downarrow \\ T/W & \xrightarrow{\cong} & G/\mathrm{Ad} \end{array}$$

Proof. (i) Let $T_\alpha = \mathrm{Ker}^0(T \xrightarrow{\alpha} U(1))$ and $G_\alpha = N^0(T_\alpha)$, the connected normalizor of T_α, $\bar{G} = G_\alpha/T_\alpha$. Then it is clear that $\bar{G}_\alpha$ is of rank one where root system $\Delta(\bar{G}_\alpha)$ consists of precisely all those roots in $\Delta(G)$ proportionate to α. However, it follows from Corollary (II.1.1) of § 1 − *A* that $\Delta(\bar{G}_\alpha) = \{\pm\alpha\}$, hence the multiplicity of α must be 1 and $k\alpha \in \Delta(G)$ iff $k = \pm 1$.

(ii) Clearly, $F(T_\alpha, \mathfrak{g}) = \mathfrak{g}_\alpha$ (the Lie algebra of $G_\alpha = N^0(T_\alpha) = Z(T_\alpha)$) and the induced action of G_α on $\mathfrak{g}_\alpha$ is the adjoint action, Ad_{G_α}, which is, effectively the rotation of $\bar{G}_\alpha$ on $\mathfrak{g}_\alpha$ with H_α (codim 3 in $\mathfrak{g}_\alpha$, codim 1 in $\mathfrak{h}$) as fixed point set. Hence, it is not difficult to see that $F(G_\alpha, \mathfrak{g}) = F(r_\alpha, F(T, \mathfrak{g})) = F(r_\alpha, \mathfrak{h}) = H_\alpha$, where r_α is the *reflection* that generates $W(G_\alpha) \cong \mathbb{Z}_2$.

(iii) Consider the Weyl group W as an (effective) transformation group on $\mathfrak{h} = F(T, \mathfrak{g})$, W contains the above reflections r_α, for each pair of roots $\pm\alpha \in \Delta(G)$. We claim that W is, in fact, generated by the above reflections $\{r_\alpha\}$. In order to show that, let W' be the subgroup generated by $\{r_\alpha, \pm\alpha \in \Delta(G)\}$. Since the set $\Delta(G)$ is invariant under W, it is clear that $(\mathfrak{h} - \bigcup\{H_\alpha\})$ is also invariant under W, consequently, W permutes its connected components, i.e., the (Weyl) chambers. Notice that W' permutes the set of chambers *simply transitively*. Hence, we need only to show that W *also* permutes the set of chambers *simply transitively* (for then, ord W = ord W' = the number of chambers). Suppose the contrary. Then there exists an element $\sigma \in W$ of order p (prime) such that $\sigma(C_0) = C_0$. Hence, it follows from a theorem of P. A. Smith (and the acyclicity of C_0) that there exists $X \in C_0$ fixed under σ and consequently, G_X is disconnected, (for $G_X^0 = T$, $G_X/G_X^0 \supseteq \{\sigma\}$) which is a contradiction to Corollary (I.8.2) of § I.4-B. Hence $W = W'$ and is generated by the reflections $\{r_\alpha\}$.

(iv) It follows directly from the principal orbit theorem that the fixed point set of a principal isotropy subgroup, $F(T, \mathfrak{g}) = \mathfrak{h}$ (resp. $F(T, G) = T$) intersects every orbit. Namely, the following inclusions induces surjections of the respective orbit spaces:

$$\begin{array}{ccc} \mathfrak{h} & \xrightarrow{\subseteq} & \mathfrak{g} \\ \downarrow & & \downarrow \\ \mathfrak{h}/W & \xrightarrow{\text{surj.}} & \mathfrak{g}/\mathrm{Ad} \end{array} \qquad \begin{array}{ccc} T & \xrightarrow{\subseteq} & G \\ \downarrow & & \downarrow \\ T/W & \xrightarrow{\text{surj.}} & G/\mathrm{Ad} \end{array}$$

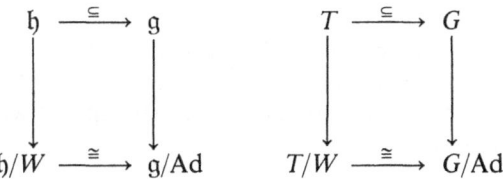

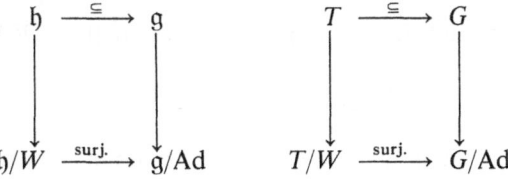

On the other hand, the injectivity of the above maps simply means that $F(T, G(X)) = N(T, G)/N(T, G_X) = W(G)/W(G_X)$ which is a direct consequence of maximal tori theorem. Finally, we remark that

$$\Delta(G_X)=\{\alpha\in\Delta(G): X\in H_\alpha\} \quad\text{and}\quad W_X=W(G_X) \quad\text{for}\quad X\in\mathfrak{h}. \quad \square$$

Definition. The orbit space $\mathfrak{g}/\mathrm{Ad}\cong\mathfrak{h}/W$ is called the *Weyl chamber* of $\mathfrak{g}$ and the orbit space $G/\mathrm{Ad}\cong T/W$ is called the *Cartan polyhedron.* In view of Theorem (II.2), it is also customary to identify the above orbit spaces with any one of their respective fundamental domains.

(D) *The Volume Function and Weyl Integration Formula*

If one equips G with a bi-invariant Riemannian metric with total volume 1, and f is a central function on G, then the Haar integral $\int_G f(g)\,dg$ can be reduced to the following *weighted* integration

$$\int_P f(t)\rho(t)dt = \frac{1}{w}\int_T f(t)\rho(t)dt \qquad (w=\mathrm{ord}(W(G)),\ \text{and } T \text{ consists of } w\,P\text{'s})$$

over the Cartan polyhedron P with $\rho(t)=q$-dim volume of $G(t)$, $q=\dim G/T$. Now let us compute the above volume function $\rho(t): T\to R$ as follows. Let us equip G/T with the homogeneous Riemannian metric such that the metric on the tangent space at the base point is the restriction of metric on $\mathfrak{g}$ to $\mathfrak{h}^\perp$. For a given $t\in T$, we have

$$\begin{array}{ccc} G/T & \xrightarrow{\iota(t)} & G(t) \\ \cup\!\!| & & \cup\!\!| \\ G_\alpha/T & \xrightarrow{\iota_\alpha(t)} & G_\alpha(t) \end{array}$$

and it is clear that $\rho(t)=\mathrm{vol}(G(t))=\mathrm{vol}(G/T)\cdot|\text{The Jacobian of } \iota(t) \text{ at } e|$. Since the tangent space at the base point $e\in G/T$, $\mathfrak{h}^\perp$, decomposes into invariant root spaces w.r.t. $d\iota(t)|e$:

$$\mathfrak{h}^\perp=\sum \mathbb{R}^2_{\pm\alpha}; \qquad \alpha\in\Delta^+$$

one has $\det(d\iota(t))_e=\prod_{\alpha\in\Delta^+}\det(d\iota_\alpha(t))_e$. In view of the computation for the rank one case, it is not difficult to see that $\det(d\iota_\alpha(t))=\text{constant}\cdot\sin^2(\pi\alpha(t))$ and hence

$$\rho(t)=c_0\cdot\det(d\iota(t))_e=c\cdot\prod_{\alpha\in\Delta^+}\sin^2(\pi\cdot\alpha(t))=c'Q(t)\cdot\overline{Q(t)}$$

where $c'=c/2^p$ and $Q(t)=\prod_{\alpha\in\Delta^+}(e^{\pi i\alpha(t)}-e^{-\pi i\alpha(t)})$.

Remark. It is useful to note that $Q(t)$ is antisymmetric w.r.t. W, i.e., $Q(\sigma(t))=\mathrm{sign}(\sigma)\cdot Q(t)$. Then it is easy to show

$$Q(t)=\sum_{\sigma\in W}\mathrm{sign}(\sigma)\,e^{2\pi i\sigma\cdot\delta(t)}$$

where $\delta=\frac{1}{2}\sum_{\alpha\in\Delta^+}\alpha$. Use the above expression of $Q(t)$, it is easy to determine the constant c' as follows

$$1=\int_G 1\cdot dg=\frac{1}{w}\int_T c'\cdot|Q(t)|^2\,dt=\frac{c'}{w}(\|Q(t)\|^2_{L^2(T)})$$

$$=\frac{c'}{w}(\textstyle\sum_{\sigma\in W}\|e^{2\pi i\sigma\delta(t)}\|^2_{L^2(T)})=c'\Rightarrow c'=1\,.$$

Hence, we have the following important formula.

Weyl Integration Formula. For a central function f on G, one has

$$\int_G f(g)\,dg=\frac{1}{w}\int_T f(t)\,|Q(t)|^2\,dt$$

where $Q(t)=\prod_{\alpha\in\Delta^+}(e^{\pi i\alpha(t)}-e^{-\pi i\alpha(t)})=\sum_{\sigma\in W}\operatorname{sign}(\sigma)e^{2\pi i\sigma\cdot\delta(t)}$ and $\delta=\frac{1}{2}\sum_{\alpha\in\Delta^+}\alpha$.

§ 2. Classification of Compact Connected Lie Groups

(A) *Cartan-Killing Form and Characterization of Compact Lie Algebras*

Definition. A (real) Lie algebra $\mathfrak{g}$ is called a *compact* Lie algebra if it is the Lie algebra of a *compact* Lie group G.

Obviously, a compact Lie algebra, $\mathfrak{g}$, has an inner product invariant under the adjoint action of G. Namely,

$$(\mathrm{Ad}(\mathrm{Exp}\,tX)\cdot Y,\ \mathrm{Ad}(\mathrm{Exp}\,tX)\cdot Z)\equiv(Y,Z),\quad \text{for all}\quad X,Y,Z\in\mathfrak{g}\,.$$

The above identity is clearly equivalent to its differentiated version;

$$\frac{d}{dt}(\mathrm{Ad}(\mathrm{Exp}\,tX)\cdot Y,\ \mathrm{Ad}(\mathrm{Exp}\,tX)\cdot Z)_{t=0}=([X,Y],Z)+(Y,[X,Z])\equiv 0\,.$$

An important, simple consequence of the above fact is the following:

Theorem (II.4). *A compact Lie algebra $\mathfrak{g}$ decomposes uniquely into the sum of its center $\mathfrak{g}_0$ and normal simple subalgebras:* $\mathfrak{g}=\mathfrak{g}_0+\sum\mathfrak{g}_j$.

Proof. It follows easily from the above identity that the perpendicular space $\mathfrak{k}^\perp$ of an arbitrary *normal* subalgebra $\mathfrak{k}$ is also a *normal* subalgebra, and $\mathfrak{g}=\mathfrak{k}+\mathfrak{k}^\perp$ (as Lie algebras), for $X\in\mathfrak{g}$, $Y\in\mathfrak{k}$, $Z\in\mathfrak{k}^\perp$

$$[X,Y]\in\mathfrak{k}\Rightarrow(Y,[X,Z])=-([X,Y],Z)=0\Rightarrow[X,Z]\in\mathfrak{k}^\perp\,.\quad\square$$

In view of the above theorem, we shall from now on assume that $\mathfrak{g}$ is itself simple, although most of the following discussions are also valid (with some obvious modifications) for general compact Lie algebras, or at least for semi-simple compact Lie algebras.

Since a linear subspace $\mathfrak{k}$ of a Lie algebra $\mathfrak{g}$ is *invariant* (w.r.t. Ad) if and only if $\mathfrak{k}$ is a *normal* subalgebra, the adjoint action of G on its Lie algebra $\mathfrak{g}$ is *irreducible* if and only if G (resp. $\mathfrak{g}$) is *simple*. Hence, for a *simple compact* Lie algebra $\mathfrak{g}$, two *invariant* symmetric bilinear forms on $\mathfrak{g}$ are proportional, i.e., $B_1(X,Y)=k\,B_2(X,Y)$. On the other hand, the following *Cartan-Killing form*

$$B(X,Y) \overset{\text{def}}{=} \operatorname{Tr}(\operatorname{ad}_X \cdot \operatorname{ad}_Y)$$

is clearly an *intrinsically defined, symmetric* bilinear form, and consequently, also *invariant* (w.r.t. inner automorphisms of $\mathfrak{g}$). Hence, the above Cartan-Killing form $B(X,Y)$ is proportional to an invariant *inner product* on $\mathfrak{g}$. Note that ad_X is anti-symmetric and the eigenvalues of ad_X are imaginary, therefore the eigenvalues of $(\operatorname{ad}_X)^2$ are negative, and $B(X,X)=\operatorname{tr}(\operatorname{ad}_X)^2<0$. Hence, $B(X,Y)$ is *negative definite*, and it is natural to consider $(X,Y)=-B(X,Y)$ as the *intrinsic* inner product on $\mathfrak{g}$. In fact, it is not difficult to prove the following characterization of compact Lie algebras:

Theorem (II.5). *A simple (or semi-simple) real Lie algebra $\mathfrak{g}$ is compact if and only if its Cartan-Killing form $B(X,Y)$ is negative definite.*

(B) *System of Simple Roots and Dynkin Diagram*

Let $\mathfrak{g}$ be a *simple compact* Lie algebra, $\mathfrak{h}$ be an arbitrarily chosen but fixed Cartan subalgebra, and $\Delta \subseteq \mathfrak{h}^*$ (the dual space of $\mathfrak{h}$) be the root system of $\mathfrak{g}$. Let $(X,Y) = -B(X,Y) = -\operatorname{tr}\operatorname{ad}_X \cdot \operatorname{ad}_X$ be the intrinsic inner product on $\mathfrak{g}$ and respectively the induced inner products on $\mathfrak{h}$ and $\mathfrak{h}^*$. Then the Weyl group W acts on $\mathfrak{h}$ (resp. on $\mathfrak{h}^*$) as an orthogonal transformation group generated by the reflections $\{r_\alpha, \pm\alpha \in \Delta\}$. Note that since W permutes the chambers simply transitively, it is convenient to choose an arbitrary but fixed chamber C_0 and then define the positivity of roots as follows:

$$\alpha>0 \Leftrightarrow \alpha(C_0)>0\,; \qquad \alpha\in\Delta \qquad [\text{positivity w.r.t. } C_0].$$

Then it is clear that Δ splits into the disjoint union of positive roots Δ^+ and negative roots Δ^-, i.e., $\Delta=\Delta^+ \cup \Delta^-$, and

$$\bar{C}_0 = \bigcap\nolimits_{\alpha\in\Delta^+} \{\mathfrak{h}_\alpha^+ = (h\in\mathfrak{h};\ \alpha(h)\geqslant 0)\}; \qquad C_0 = \bigcap\nolimits_{\alpha\in\Delta^+} \{\mathring{\mathfrak{h}}_\alpha^+\}.$$

Definition. Let $\pi \in \Delta^+$ be the subset of those positive roots whose hyperplanes have codimension one intersections with $\bar{C}_0$. Then it is clear that π is the minimal subset of Δ^+ with $\bar{C}_0 = \bigcap_{\alpha\in\pi}\{\mathfrak{h}_\alpha^+\}$. π is called the *system of simple roots* (w.r.t. C_0).

Theorem (II.6). (i) *Let $\alpha, \beta \in \Delta$ and $p, q \geqslant 0$ be the respective largest integers such that $(\beta + p\alpha)$, $(\beta - q\alpha) \in \Delta$. Then $(\beta + j\alpha) \in \Delta$ for $-q \leqslant j \leqslant p$ and $2(\beta,\alpha)/(\alpha,\alpha) = (q-p)$.*

(ii) *Let $\pi \subset \Delta^+$ be the system of simple roots. Then $\alpha_i \neq \alpha_j \in \pi$ implies $(\alpha_i, \alpha_j) \leqslant 0$ and π forms a basis of $\mathfrak{h}^*$ such that every roots $\beta \in \Delta$ is an integral linear combination of simple roots with uniform sign, i.e.,*

$$\beta = \sum_{\alpha_j \in \pi} k_j \alpha_j; \quad k_j \in \mathbb{Z} \text{ and } \geqslant 0 \text{ if } \beta \in \Delta^+ \text{ (resp. } \leqslant 0 \text{ if } \beta \in \Delta^-).$$

(iii) *Let $\mathfrak{g}_1, \mathfrak{g}_2$ be two simple compact Lie algebra Δ_1, Δ_2 and π_1, π_2 be respectively their root systems and systems of simple roots. Then $\mathfrak{g}_1 \cong \mathfrak{g}_2 \Rightarrow \pi_1$ and π_2 isometric, and an isometry of π_1, π_2 can be uniquely extended to an isometry of Δ_1, Δ_2.*

Proof. (i) Since Δ is invariant under the Weyl group W, it is obvious that $r_\alpha(\beta + p\alpha) = (\beta + p\alpha) - 2\dfrac{(\beta + p\alpha, \alpha)}{(\alpha, \alpha)} \cdot \alpha = (\beta - q\alpha)$. Hence, one has $2(\beta, \alpha)/(\alpha, \alpha) = (q - p)$. The fact that $(\beta + j\alpha) \in \Delta$ for $-q \leqslant j \leqslant p$ follows directly from Theorem (II.1) of § 1-A applies to $\mathrm{Ad}_G | G_\alpha$.

(ii) Observe that if a positive root α can be decomposed into the sum of two other positive roots, $\alpha = \alpha_1 + \alpha_2$, then the condition $\alpha(h) \geqslant 0$ is already implied by the conditions $\alpha_1(h) \geqslant 0$ and $\alpha_2(h) \geqslant 0$, and hence can be omitted from the above expression of $\bar{C}_0$ as intersection of half spaces. Therefore, it is easy to see that $\pi \subseteq \Delta^+$ are exactly those *indecomposable* positive roots. Let $\alpha_i \neq \alpha_j \in \pi$ be two simple roots. Then $(\alpha_i - \alpha_j) \notin \Delta$, for otherwise, either $(\alpha_i - \alpha_j) \in \Delta^+ \Rightarrow \alpha_i = (\alpha_i - \alpha_j) + \alpha_j$ is decomposable or $(\alpha_j - \alpha_i) \in \Delta^+$ and $\alpha_j = (\alpha_j - \alpha_i) + \alpha_i$ is decomposable. Hence, it follows from (i) that $q = 0$ and $2(\alpha_i, \alpha_j)/(\alpha_j, \alpha_j) = (q - p) = -p \leqslant 0$, i.e., $(\alpha_i, \alpha_j) \leqslant 0$. Then, it is a simple fact of linear algebra that positivity of α_i and $(\alpha_i, \alpha_j) \leqslant 0$ for all $1 \leqslant i \leqslant j \leqslant r \Rightarrow \pi = \{\alpha_1, \ldots, \alpha_r\}$ linearly independent. Therefore every positive (resp. negative) root $\beta \in \Delta^+$ can be expressed *uniquely* as linear combination of $\alpha_j \in \pi$ with non-negative (resp. non-positive) integral coefficients. The fact that π spans $\mathfrak{h}^*$ follows easily from the fact that $\mathfrak{g}$ has no center.

(iii) Since all Cartan subalgebras of a compact Lie algebra are conjugate to each other, one may modify the given isomorphism $\iota: \mathfrak{g}_1 \to \mathfrak{g}_2$ by a suitable inner automorphism so that $\iota(\mathfrak{h}_1) = \mathfrak{h}_2$. Hence it follows directly from the definition of root system and the *intrinsic* inner product that ι induces an isometry of Δ_1 onto Δ_2. Furthermore, since W acts simply transitively on the set of chambers and the choice of a system of simple roots, π, is in $1-1$ correspondence with the choice of a chamber, it is clear that W also *permutes simply transitively* among *different systems of simple roots.* Hence, after a suitable modification by a conjugation of an element of W, we have the induced isometry maps π_1 onto π_2.

Finally, it is an easy consequence of (i) that Δ is completely determined by the metric property of π, and hence an isometry of π_1 onto π_2 extends uniquely to an isometry of Δ_1 onto Δ_2. □

Dynkin diagram.

Observe that $\dfrac{2(\alpha_i, \alpha_j)}{(\alpha_i, \alpha_i)} \cdot \dfrac{2(\alpha_i, \alpha_j)}{(\alpha_j, \alpha_j)} < 4$ implies that the integer $2(\alpha_i, \alpha_j)/(\alpha_i, \alpha_i)$ is either 0, -1, -2 or -3, which geometrically corresponds to the cases that the angle between α_i and α_j is $90°$, $120°$, $135°$, or $150°$ respectively. Therefore, it is convienent to record the metric property of the system of simple roots π in terms of the following diagram:

Symbolically, we represent each simple root by a point and we join two points by a single, or double, or triple bond if the angle between the respective simple roots is 120°, or 135°, or 150°. Moreover, in the case of double or triple bond, i.e., $2(\alpha_i, \alpha_j)/(\alpha_i, \alpha_i) = -2$, or -3 and $2(\alpha_i, \alpha_j)/(\alpha_j, \alpha_j) = -1$, the two simple roots are not of equal length. Hence it is natural to use directed bond ($\Rightarrow$ or $\Rrightarrow$) to indicate which one is longer than the other. Such a diagram is called the *Dynkin diagram* of the system of simple roots π (or of the Lie algebra $\mathfrak{g}$).

Proposition. *The Dynkin diagram of a simple compact Lie algebra is connected. In general, there is a 1−1 correspondence between the connected components of its Dynkin diagram and the simple normal subalgebras of a compact Lie algebra* $\mathfrak{g}$.

Proof. Suppose π' is a connected component of π and $\pi = \pi' + \pi''$. Then it follows from definition that $\pi' \perp \pi''$. Hence, by the above theorem, $(\alpha', \alpha'') = 0$ if α', α'' are respectively linear combinations of π', π'', and therefore $\alpha' + \alpha'' \notin \Delta$ (for $2(\alpha', \alpha'')/(\alpha', \alpha') = (q-p) = -p = 0$) which in turn implies $[X_{\alpha'}, X_{\alpha''}] = 0$. Then, it is easy to show that the subalgebra generated by $X_{\alpha'}$, $\alpha' \in \langle \pi' \rangle$ is a normal subalgebra and the proposition follows. □

A connected Dynkin diagram is called *geometrically feasible* if there exists a set of vectors with the metric property indicated by the given diagram. Of course, for the purpose of such a purely geometric consideration, only the angles are essential and it is not difficult to see that a necessary and sufficient condition for a set of vectors $\{\alpha_1, \ldots, \alpha_r\}$ with preassigned angles is *geometrically feasible* is that $|\sum t_j \alpha_j|^2 = (\sum t_j \alpha_j, \sum t_j \alpha_j) \geq 0$ for any $t_j \in \mathbb{R}$, (and $=0$ only when $\sum t_j \alpha_j = 0$). Hence, it is rather elementary to prove the following:

Theorem (II.7). *There are only the following geometrically feasible connected Dynkin diagram:*

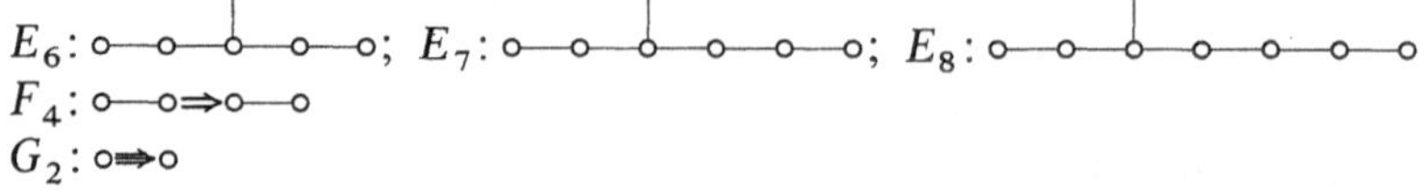

We refer the reader to § 5, Ch. IV of Jacobson's Lie Algebra for a standard proof of the above elementary theorem.

(C) *Chevalley Basis and Classification Theorem*

In the case of compact Lie algebras, the classification theorem of Cartan-Killing can be simply stated as follows:

Theorem (II.8) (Cartan-Killing). *The map of assigning a semi-simple compact Lie algebra $\mathfrak{g}$ to its Dynkin diagram $D(\mathfrak{g})$ is a bijection between the set of isomorphic classes of semi-simple compact Lie algebras and the set of all geometrically feasible Dynkin diagrams.* (cf. Theorem (II.6).

We shall prove the above theorem as follows:

Let $\mathfrak{g}$ be a semi-simple compact Lie algebra $\mathfrak{g}_c = \mathfrak{g} \otimes \mathbb{C}$ be its complexification. Let $\mathfrak{h}$ be a Cartan subalgebra and Δ be the root system of $\mathfrak{g}$. Then, by definition, we have the following decomposition of $\mathfrak{g}_c$ w.r.t. Ad_T (or $\mathrm{ad}\,\mathfrak{h}$):

$$\mathfrak{g}_c = \mathfrak{h} \otimes \mathbb{C} + \textstyle\sum_{\alpha \in \Delta} \mathbb{C}_\alpha ,$$

where $\mathbb{C}_\alpha = \{X \in \mathfrak{g}_c;\, [H,X] = i\alpha(H)\cdot X \text{ for } H \in \mathfrak{h}\}$ and are one-dimensional. If one restrict the adjoint action to those subgroups, $\{G_\alpha; \alpha \in \Delta^+\}$, next to the maximal torus T, then the above root-space decomposition are strung into invariant subspaces of Ad_{G_α} (resp. $\mathrm{ad}_{\mathfrak{g}_\alpha}$) as follows:

$$\mathfrak{g}_c = \mathfrak{g}_\alpha \otimes \mathbb{C} + \textstyle\sum (\sum_{j=-q}^{p} \mathbb{C}_{\beta + j\alpha}) .$$

It is, then, straightforward to verify the following properties of root-spaces decomposition:

(i) For $\alpha, \beta \in \Delta, [\mathbb{C}_\alpha, \mathbb{C}_\beta] = \mathbb{C}_{\alpha+\beta}$ if we set $\mathbb{C}_{\alpha+\beta} = 0$ for the case $\alpha + \beta \notin \Delta$.

Proof. $[H,[X_\alpha, X_\beta]] = [[H,X_\alpha],X_\beta] + [X_\alpha,[H,X_\beta]] = i(\alpha+\beta)(H)\cdot[X_\alpha, X_\beta]$, $H \in \mathfrak{h}$, $X_\alpha \in \mathbb{C}_\alpha$, $X_\beta \in \mathbb{C}_\beta$.

and the fact that $\alpha + \beta \in \Delta \Rightarrow [X_\alpha, X_\beta] \neq 0$ follows from the fact that $(\sum_{j=-q}^{p} \mathbb{C}_{\beta+j\alpha})$ is irreducible w.r.t. $\mathrm{ad}_{\mathfrak{g}_\alpha}$.

(ii) For each root $\alpha \in \Delta$, let $H'_\alpha \in \mathfrak{h}$ be such that $(H'_\alpha, H) = \alpha(H)$ for all $H \in \mathfrak{h}$ and $H_\alpha = 2H'_\alpha/(\alpha,\alpha)$. Then each H_α is an *integral* linear combination of $H_i = H_{\alpha_i}$ where $\{\alpha_1, \ldots, \alpha_r\} = \pi$ is a system of simple roots.

Proof. Write r_j for $r_{\alpha_j} \in W$. Then

$$r_j H'_i = H'_i - \frac{2(\alpha_i, \alpha_j)}{(\alpha_j, \alpha_j)} H'_j ,$$

and hence

$$r_j H_i = \frac{2}{(\alpha_i,\alpha_i)} r_j H'_i = \frac{2}{(\alpha_i,\alpha_i)} \cdot H'_i - \frac{2}{(\alpha_i,\alpha_i)} \frac{2(\alpha_i,\alpha_j)}{(\alpha_j,\alpha_j)} H'_j = H_i - \frac{2(\alpha_i,\alpha_j)}{(\alpha_i,\alpha_i)} H_j = H_{r_j\alpha_i} .$$

since $\{r_j\}$ generates W, and $\alpha = w\cdot\alpha_i$ for suitable $w \in W$ and $\alpha_i \in \pi$, (ii) follows.

(iii) By looking at $\mathfrak{g}_\alpha \otimes \mathbb{C} = \mathfrak{h}_\alpha \otimes \mathbb{C} + \bar{\mathfrak{g}}_\alpha \otimes \mathbb{C} = \mathfrak{h}_\alpha \otimes \mathbb{C} + (\{H_\alpha\} \otimes \mathbb{C} + \mathbb{C}_\alpha + \mathbb{C}_{-\alpha})$, it is obvious that $[X_\alpha, X_{-\alpha}] = \lambda \cdot H_\alpha \neq 0$ for $X_\alpha, X_{-\alpha}$ non-zero vectors of $\mathbb{C}_\alpha$ and $\mathbb{C}_{-\alpha}$ respectively. Moreover, the following identity

$$([X_\alpha, X_{-\alpha}], H) = -(X_{-\alpha}, [X_\alpha, H]) = i\alpha(H)(X_{-\alpha}, X_\alpha), \quad H \in \mathfrak{h}$$

implies that $[X_\alpha, X_{-\alpha}] = i(X_\alpha, X_{-\alpha})\cdot H'_\alpha = \frac{1}{2}(X_\alpha, X_{-\alpha})(\alpha,\alpha)\, iH_\alpha$, $(X_\alpha, X_{-\alpha}) \neq 0$. In fact, it is not difficult to show that there exist $X_\alpha, X_{-\alpha}$ such that $\bar{X}_\alpha = X_{-\alpha}$ and $(X_\alpha, X_{-\alpha}) = 2/|\alpha|^2$, i.e., $[X_\alpha, X_{-\alpha}] = iH_\alpha$. [Two such pairs differ by a factor of $e^{i\theta}$, i.e., $\{e^{i\theta}X_\alpha, e^{-i\theta}X_{-\alpha}\}$.]

(iv) Let $\{X_\alpha, \alpha\in\Delta\}$ be so chosen that $\bar{X}_\alpha = X_{-\alpha}$ and $[X_\alpha, X_{-\alpha}] = iH_\alpha$. Define $N_{\alpha,\beta}$ by $[X_\alpha, X_\beta] = N_{\alpha,\beta} X_{\alpha+\beta}$ if $(\alpha+\beta)\in\Delta$ and $N_{\alpha,\beta}=0$ if $(\alpha+\beta)\notin\Delta$. Then one has the following properties of the structural coefficients $N_{\alpha,\beta}$.

(a) $N_{-\alpha,-\beta} = \bar{N}_{\alpha,\beta}$ and $N_{\alpha,\beta} = -N_{\beta,\alpha}$: Obvious from definition.

(b) For a triangle of roots $\alpha+\beta+\gamma=0$, one has $\dfrac{N_{\alpha,\beta}}{|\gamma|^2} = \dfrac{N_{\beta,\gamma}}{|\alpha|^2} = \dfrac{N_{\gamma,\alpha}}{|\beta|^2}$.

Proof.
$$2\frac{N_{\alpha\beta}}{|\gamma|^2} = (N_{\alpha\beta}X_{-\gamma}, X_\gamma) = ([X_\alpha, X_\beta], X_\gamma)$$
$$= -(X_\beta, [X_\alpha X_\gamma]) = N_{\gamma\alpha}(X_\beta, X_{-\beta}) = 2\frac{N_{\gamma\alpha}}{|\beta|^2}.$$

(c) Suppose $\alpha,\beta,\gamma,\delta,\in\Delta$ and $\alpha+\beta+\gamma+\delta=0$ but *no* two are proportional. Then $[X_\alpha,[X_\beta,X_\gamma]] = N_{\beta\gamma}[X_\alpha, X_{\beta+\gamma}] = N_{\beta\gamma}\cdot N_{\alpha,\beta+\gamma}X_{-\delta} = N_{\beta\gamma}N_{\delta\alpha}\dfrac{|\delta|^2}{|\beta+\gamma|^2}\cdot X_{-\delta}$.

Hence, it follows from the Jacobi identity that
$$N_{\alpha\beta}N_{\gamma\delta}|\alpha+\beta|^{-2} + N_{\beta\gamma}N_{\alpha\delta}\cdot|\beta+\gamma|^{-2} + N_{\gamma\alpha}\cdot N_{\beta\delta}\cdot|\gamma+\alpha|^{-2} = 0.$$

(d) $|N_{\alpha\beta}|^2 = p(q+1)\dfrac{|\alpha+\beta|^2}{|\beta|^2} = (q+1)^2$;
(where $\alpha,\beta\in\Delta$ and $\beta+j\alpha\in\Delta$, $-q\leqslant j\leqslant p\geqslant 1$).

Proof.
$$[X_{-\alpha}[X_\alpha, X_\beta]] = N_{\alpha\beta}[X_{-\alpha}X_{\alpha+\beta}]$$
$$= N_{\alpha\beta}\cdot N_{-\alpha,\alpha+\beta}X_\beta \qquad \text{apply (b) to } (-\alpha),(-\beta),\ (\alpha+\beta)]$$
$$= N_{\alpha\beta}\cdot N_{-\beta,-\alpha}\frac{|\beta|^2}{|\alpha+\beta|^2}\cdot X_\beta$$
$$= -|N_{\alpha\beta}|^2\cdot\frac{|\beta|^2}{|\alpha+\beta|^2}X_\beta \quad \text{[by (a)]}.$$

On the other hand, it is a simple fact of the G_α (or rather, $\mathfrak{g}_\alpha$) representation on $(\sum \mathbb{C}_{\beta+j\alpha})$ that $[X_{-\alpha}[X_\alpha, Y_\beta]] = -p(q+1)Y_\beta$ for any $Y_\beta\in\mathbb{C}_\beta$. Hence, one has
$$|N_{\alpha\beta}|^2 = p(q+1)\frac{|\alpha+\beta|^2}{|\beta|^2}.$$

Furthermore, let $T_{\alpha\beta} = T_\alpha\cap T_\beta$, $G_{\alpha\beta} = N^0(T_{\alpha\beta})$ and $\bar{G}_{\alpha\beta} = G_{\alpha\beta}/T_{\alpha\beta}$. It is clear that $\bar{G}_{\alpha\beta}$ is of rank 2 and $(\bar{G}_{\alpha\beta})$ consists of all those roots of Δ which are linear combination of α,β. Since $\alpha+\beta\in\Delta(\bar{G}_{\alpha\beta})$, the Dynkin diagram of $\bar{G}$ must be connected and hence either o—o, or o⇒o, *or* o⇛o. Then it is a simple matter (though a little tedious) to check in the above three cases that $p(q+1)\frac{|\alpha+\beta|^2}{|\beta|^2}=(q+1)^2$.

Theorem (II.8′) (Chevalley). *Let* $\mathfrak{g}$ *be a compact semi-simple Lie algebra and* $\mathfrak{g}_c=\mathfrak{g}\otimes\mathbb{C}$ *be its complexification;* $\mathfrak{g}_c=\mathfrak{h}\otimes\mathbb{C}+\sum_{\alpha\in\Delta}\mathbb{C}_\alpha$. *Then, it is possible to choose* $X_\alpha\in\mathbb{C}_\alpha$ *such that*

$$\bar{X}_\alpha=X_{-\alpha};\ [X_\alpha,X_{-\alpha}]=iH_\alpha=\text{integral linear combination of } iH_{\alpha_j},$$
$$[X_\alpha,X_\beta]=\pm(q+1)X_{\alpha+\beta}.$$

Hence the above $\{X_\alpha,\alpha\in\Delta\}$ *together with* $\{iH_{\alpha_j};\alpha_j\in\pi\}$ *form a basis of* $\mathfrak{g}_c$ *such that all the structural constants are integral. It is called the Chevalley basis of* $\mathfrak{g}_{\mathbb{C}}$.

Remarks (i) $\{H_{\alpha_j},\alpha_j\in\pi\}\cup\{(X_\alpha+X_{-\alpha}),\ i(X_\alpha-X_{-\alpha});\ \alpha\in\Delta^+\}$ forms a basis of $\mathfrak{g}$.

(ii) The above theorem clearly implies the "if part" of the classification theorem, namely, $\Delta_1\cong\Delta_2\Rightarrow\mathfrak{g}_1\cong\mathfrak{g}_2$ (or resp. $\mathfrak{g}_{1c}\cong\mathfrak{g}_{2c}$).

(iii) Since the existence of Lie algebras of the classical types, i.e., A_n, B_n, C_n, D_n, is a well known fact, one need only to show the existence of a simple Lie algebra; for each of the five exceptional types. In view of the above explicit basis and structural constants, it is a matter of straightforward verification.

Proof of the Chevalley theorem. Since $|N_{\alpha\beta}|^2=(q+1)^2$, one need only to show that it is possible to choose X_α so that $N_{\alpha,\beta}$ are all *real* numbers. Note that two pairs $\{X_\alpha,X_{-\alpha}\}$ and $\{X'_\alpha,X'_{-\alpha}\}$ with $\bar{X}_\alpha=X_{-\alpha}$, $\bar{X}'_\alpha=X'_{-\alpha}$ and $(X_\alpha,X_{-\alpha})=(X'_\alpha,X'_{-\alpha})=2/|\alpha|^2$ differ by a factor of $e^{i\theta}$, i.e., $X'_\alpha=e^{i\theta}\cdot X_\alpha$ and $X'_{-\alpha}=e^{-i\theta}X_{-\alpha}$, It is natural to begin with an arbitrary basis $\{X_\alpha,X_{-\alpha};\alpha\in\Delta^+\}$ with $\bar{X}_\alpha=X_{-\alpha}$ and $(X_\alpha,X_{-\alpha})=2/|\alpha|^2$ and then inductively adjust each pair by suitable factor of $e^{i\theta}$ to make $N_{\alpha,\beta}$ all real. Let $\Delta_\rho=\{\alpha\in\Delta;-\rho<\alpha<\rho\},\rho\in\Delta^+$. We may assume that $N_{\alpha,\beta}\in\mathbb{R}$ for $\alpha,\beta,(\alpha+\beta)\in\Delta_\rho$ and proceed to prove the induction step that $N_{\alpha,\beta}\in\mathbb{R}$ for $\alpha,\beta,(\alpha+\beta)\in\Delta_\rho\cup\{\pm\rho\}$. If ρ can not be expressed as the sum of two vectors of Δ_ρ, then we don't have to adjust $\{X_\rho,X_{-\rho}\}$, Otherwise, let $\rho=\alpha+\beta$ be such an expression with smallest α. We simply adjust $\{X_\rho,X_{-\rho}\}$ so that $N_{\alpha,\beta}$ is real. (In fact, there are exact two ways by making $N_{\alpha\beta}>0$ or <0 respectively). Suppose $\lambda+\mu=\alpha+\beta=\rho$ is another such expression. Then $\alpha+\beta+(-\lambda)+(-\mu)=0$ and it follows from (c) of (iv) that $N_{\lambda\mu}$ is also real. This completes the induction step and the theorem follows by induction. □

(D) *A Theorem of Weyl and the Determination of* $Z(G)$ *for Simple Connected* G

Theorem (II.9) (Weyl). *Let* G *be a semi-simple compact connected Lie group. Then the simply connected (or universal) covering group* $\tilde{G}$ *of* G *is also compact.*

Proof. Suppose the contrary. Then $\ker(\tilde{G}\to G)\subseteq Z(\tilde{G})$ is an infinite discrete abelian subgroup of $\tilde{G}$. Hence it is easy to see that there are *compact* covering groups G_1; $\tilde{G}\to G_1\to G$, with $Z(G_1)$ of arbitrary large finite order, which clearly contradicts the fact that $\operatorname{ord}(Z(G_1))\leqslant$ the number of vertices in the Cartan polyhedron of G_1 (obviously bounded). □

Finally, for the sake of reference, we list the Dynkin diagram of the Cartan polydera together with the centers of those simple, compact, simply connected Lie groups as follows:

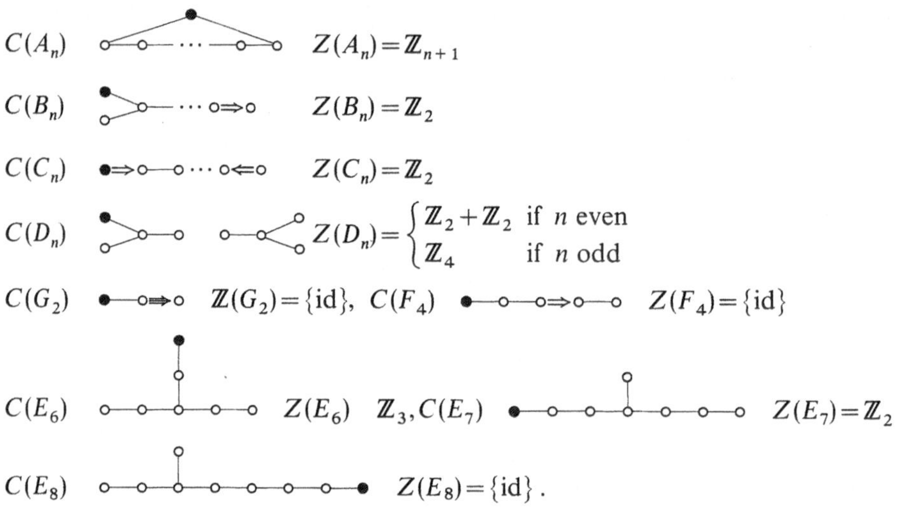

$C(A_n)$ $Z(A_n)=\mathbb{Z}_{n+1}$

$C(B_n)$ $Z(B_n)=\mathbb{Z}_2$

$C(C_n)$ $Z(C_n)=\mathbb{Z}_2$

$C(D_n)$ $Z(D_n)=\begin{cases}\mathbb{Z}_2+\mathbb{Z}_2 & \text{if } n \text{ even}\\ \mathbb{Z}_4 & \text{if } n \text{ odd}\end{cases}$

$C(G_2)$ $\mathbb{Z}(G_2)=\{\mathrm{id}\}$, $C(F_4)$ $Z(F_4)=\{\mathrm{id}\}$

$C(E_6)$ $Z(E_6)$ $\mathbb{Z}_3$, $C(E_7)$ $Z(E_7)=\mathbb{Z}_2$

$C(E_8)$ $Z(E_8)=\{\mathrm{id}\}$.

Remark. (i) In the above diagram, each dot represesents a "wall" of the Cartan polyhedron, they are respectively $\alpha_j(H)\geqslant 0, \alpha_j\in\pi$ and $\beta(H)\leqslant 1$ where β is the *highest root* which is represented by the dark dot.

(ii) Let x be a vertex of the Cartan polyhedron and $\hat{x}$ be the opposite wall of x. Then the Dynkin diagram $D(G_x)$ of the centralizor of x, G_x, is exactly the one obtained by removing the dot of $\hat{x}$ from the above diagram of Cartan polyhedron. Hence, $Z(G)$ is in $1-1$ correspondence with those vertices with $C(G)-\{\hat{x}\}=D(G)$.

§ 3. Classification of Irreducible Representations

(A) **Classification Theorem (II.10)** (Cartan-Weyl). *Let G be a simply connected, semi-simple compact Lie group, $\mathfrak{g}$ be its Lie algebra, $\mathfrak{h}$ be a Cartan subalgebra and W be the Weyl group of G. Also let Δ be the root system and π be the system of simple roots (w.r.t. a fixed ordering). Let ψ be an irreducible complex representation of G and $\Omega(\psi)$ be the weight system of ψ. We shall denote the largest weight vector in $\Omega(\psi)$ (w.r.t. the fixed ordering) by Λ_ψ and call it the highest weight of ψ. Then,*

(i) *The multiplicity of Λ_ψ is one, and any two irreducible complex representations ψ, φ are equivalent iff their highest weights are the same, i.e., $\psi\sim\varphi\Leftrightarrow\Lambda_\psi=\Lambda\varphi$.*

(ii) *The character of ψ can be given in terms of Λ_ψ by the following formula of Weyl:*

$$\chi_\psi(t)=\frac{\sum_{\sigma\in W}\det(\sigma)e^{2\pi i\sigma(\Lambda_\psi+\delta)(t)}}{\sum_{\sigma\in W}\det(\sigma)e^{2\pi i\sigma\delta(t)}},\quad \textit{where}\quad \delta=\tfrac{1}{2}\sum_{\alpha\in\Delta^+}\alpha.$$

(iii) *A vector $\Lambda\in\mathfrak{h}^*$ can be realized as the highest weight of an irreducible complex representation iff $2(\Lambda,\alpha_j)/(\alpha_j,\alpha_j)=q_j$ are non-negative integers for $\alpha_j\in\pi$.*

Proof. Let $\chi_\psi(g)$ be the character function of ψ and $\chi_\psi(t)$ be its restriction to the maximal torus T (with $\mathfrak{h}$ as its Lie algebra). Let $m(w)$ be the multiplicity of w in $\Omega(\psi)$. Then, by definition, $\chi_\psi(t)=\sum m(w)e^{2\pi i w(t)}$.

In view of the Weyl integration formula (cf. § 1-D), it is natural to try to determine the function $\chi_\psi(t)\cdot Q(t)=\chi_\psi(t)\cdot(\sum_{\sigma\in W}\det\sigma\, e^{2\pi i\sigma\delta(t)})$. Note that $\chi_\psi(t)$ is symmetric and $Q(t)$ is anti-symmetric and hence $\chi_\psi(t)\cdot Q(t)$ is anti-symmetric (w.r.t. to W-action). If one expands an anti-symmetric function f in terms of linear combination of L^2-basis of $L^2(T)$ consists of representation functions, it is easy to see that

$$f=\sum_{v\in C_0} c_v\cdot(\sum_{\sigma\in W}\det(\sigma)e^{2\pi i\sigma\cdot v(t)})$$

where v runs through weight vectors in the positive Weyl chamber C_0. Hence,

$$\begin{aligned}\chi_\psi(t)\cdot Q(t) &= \{m(\Lambda_\psi)\cdot e^{2\pi i\Lambda_\psi(t)}+\cdots\}\cdot\{\sum_{\sigma\in W}\det(\sigma)e^{2\pi i\sigma\cdot\delta(t)}\}\\ &= m(\Lambda_\psi)\cdot\sum_{\sigma\in W}\det(\sigma)\cdot e^{2\pi i\sigma(\Lambda_\psi+\delta)(t)}+\text{possible more terms.}\end{aligned}$$

Now the irreducibility of ψ (cf. Theorem (I.3′), § 1-C, Ch. I) implies that

$$\begin{aligned}1 &= \int_G \chi_\psi(g)\cdot\overline{\chi_\psi(g)}\,dg = \frac{1}{w}\int_T|\chi_\psi(t)Q(t)|^2\,dt = \frac{1}{w}\|\chi_\psi(t)\cdot Q(t)\|^2_{L^2(T)}\\ &= \frac{1}{w}\|m(\Lambda_\psi)\cdot\sum_{\sigma\in W}\det(\sigma)e^{2\pi i\sigma(\Lambda_\psi+\delta)(t)}+\cdots\|_{L^2(T)} \quad\text{(by Schur orthogonality)}\\ &= \frac{1}{w}\{m(\Lambda_\psi)^2\cdot w+\cdots\}\geqslant m(\Lambda_\psi)^2\,.\end{aligned}$$

Hence, one must have $m(\Lambda_\psi)=1$ and $\chi_\psi(t)\cdot Q(t)=\sum_{\sigma\in W}\det(\sigma)e^{2\pi i\sigma(\Lambda_\psi+\delta)(t)}$ which is exactly the Weyl character formula. Since the character is a *complete* invariant and the above Weyl character formula gives an explicit expression of $\chi_\psi(t)$ in turms of the highest weight Λ_ψ, it is obvious that $\psi\sim\varphi\Leftrightarrow\Lambda_\psi=\Lambda_\varphi$. Therefore, (i) and (ii) are completely proved; (iii) is a direct consequence of the completeness theorem of Peter-Weyl. □

Corollary (II.10.1). $\dim\psi=\prod_{\alpha\in\Delta^+}\dfrac{(\Lambda_\psi+\delta,\alpha)}{(\delta,\alpha)}$.

Proof. Observe that $\dim\psi=\chi_\psi(\mathrm{id})=\chi_\psi(0)$. However, the above formula of $\chi_\psi(t)$ reduces to a meaningless form of 0/0 if one simply substitute zero into it. Hence, we shall instead use the formula to compute

$$\dim\psi=\lim_{t\to 0}\chi_\psi(t).$$

For this purpose, it is convient to identify $\mathfrak{h}^*$ with $\mathfrak{h}$ via the inner product and rewrite the Weyl formula as

$$\chi(t)=\frac{\sum\det(\sigma)e^{2\pi i\langle\sigma(\Lambda+\delta),t\rangle}}{\sum\det(\sigma)e^{2\pi i\langle\sigma\delta,t\rangle}}\,.$$

Notice that

$$\begin{aligned}\sum \det(\sigma)\, e^{2\pi i \langle \sigma(\Lambda+\delta), s\delta\rangle} &= \sum \det(\sigma)\, e^{2\pi i \langle \sigma\cdot\delta, s(\Lambda+\delta)\rangle} \\ &= Q(s\cdot(\Lambda+\delta)) = \prod\nolimits_{\alpha\in\Delta^+} 2i\sin(\pi\langle\alpha, s(\Lambda+\delta)\rangle).\end{aligned}$$

(Cf. § 1-D) Hence

$$\begin{aligned}\dim\psi &= \lim\nolimits_{s\to 0}\chi(s\cdot\delta) = \lim\nolimits_{s\to 0}\frac{Q(s\cdot(\Lambda+\delta))}{Q(s\cdot\delta)} \\ &= \prod\nolimits_{\alpha\in\Delta^+}\lim\nolimits_{s\to 0}\frac{\sin\pi\langle\alpha, s(\Lambda+\delta)\rangle}{\sin\pi\langle\alpha, s\cdot\delta\rangle} = \prod\nolimits_{\alpha\in\Delta^+}\frac{\langle\alpha,\Lambda+\delta\rangle}{\langle\alpha,\delta\rangle}. \quad \square\end{aligned}$$

Chapter III. An Equivariant Cohomology Theory Related to Fibre Bundle Theory

In the application of cohomology theory to the study of topological transformation groups, a natural and convenient formalism is to define an *equivariant cohomology theory* for the category of G-spaces which *effectively* reflects the cohomological behavior of both *the space* and *the G-action.* Following an idea of A. Borel [cf. B 10], we shall define the *equivariant cohomology* of a G-space X to be the *ordinary cohomology* of the *total space* X_G of the *universal* bundle, $X \to X_G \to B_G$, with the given G-space X as its typical fibre, namely

$$H_G^*(X) \overset{\text{def}}{=\!=} H^*(X_G), \quad \text{where} \quad X_G = E_G \times_G X = (E_G \times X)/G.$$

The rationale of adopting the above equivariant cohomology theory in terms of the universal bundle construction is roughly the following:

(i) Intuitively and heuristically, the complexity of the G-action on X will be reflected in the complexity of the associated universal bundle $X \to X_G \to B_G$, e.g., the associated universal bundle is trivial if and only if the G-action on X is trivial. And the classical obstruction theory, especially the characteristic classes theory, clearly demonstrates that cohomology theory can then be used to detect the complexity of $X_G \to B_G$, which, in turn, reflects the complexity of the G-action itself.

(ii) Technically, it is not difficult to see that such an equivariant cohomology theory not only possesses convenient formal properties but is also effectively computable.

§ 1. The Construction of $H_G^*(X)$ and its Formal Properties

(A) *The Construction of A. Borel*

Let X be a given G-space and $E_G \to B_G$ be the universal G-bundle. Then the total space X_G of the associated universal bundle with X as fibre may be regarded as: the orbit space of $E_G \times X$

$$X_G = E_G \times_G X = (E_G \times X)/G$$

where the G-action is given by $g\cdot(e,x)=(eg^{-1},gx)$. Since the two projections are obviously equivariant, one has the following commutative diagram:

$$\begin{array}{ccccc} E_G & \longleftarrow & E_G\times X & \longrightarrow & X \\ \big\downarrow & & \big\downarrow & & \big\downarrow \\ B_G & \xleftarrow{\pi_1} & X_G & \xrightarrow{\pi_2} & X/G \end{array}$$

Next suppose that Y is a K-space, $h\colon G\to K$ is a homomorphism and $f\colon X\to Y$ is an h-equivariant map, i.e., $f(g\cdot x)=h(g)\cdot f(x)$. Then, it is easy to check that, correspondingly, there is the following commutative diagram:

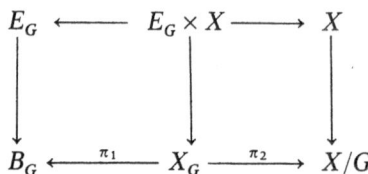

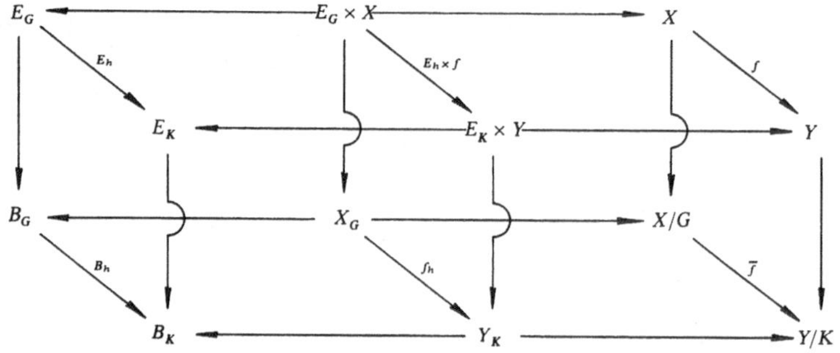

Hence, it is clear that the above construction is functorial and it follows readily that its composition with the ordinary cohomology theory will yield an *equivariant cohomology theory.*

Definition. Let $\mathscr{E}$ be the equivariant category of spaces with topological actions and equivariant maps. Then the following functor

$$(G\text{-space } X)\longmapsto H_G^*(X)=H^*(X_G),$$
$$(h\text{-equivariant map}: X\xrightarrow{f} Y)\longmapsto f_h^*: H_K^*(Y)\longrightarrow H_G^*(X)$$

is called the equivariant cohomology functor.

(B) *The Coefficient System of $H_G^*(\cdot)$*

For a *fixed* group G, it is obvious that all G-spaces and G-equivariant maps form a sub-category $\mathscr{E}_G$, and the restriction of the above theory to $\mathscr{E}_G$ will be simply called the H_G^*-theory. Since the set of homogeneous G-spaces, $\{G/H,\ H$ closed subgroups of $G\}$, are exactly those atomic G-spaces, their values of H_G^*-functor can be considered as the *coefficient system* of H_G^*-theory. Observe that

$$(G/H)_G=E_G\times_G(G/H)=(E_G\times_G G)/H=(E_G)/H=B_H\,,$$

and equivariant maps between homogeneous G-spaces are given by

$$G/H \to G/K \quad \text{for pairs of subgroups} \quad H \subseteq K.$$

Therefore, the coefficient system of the H_G^*-theory consists of the following algebras and morphisms:

$$H_G^*(G/H) = H^*(B_H) \quad \text{for each closed subgroup } H \subseteq G,$$
$$H_G^*(G/K) = H^*(B_K) \to H^*(B_H) = H_G^*(G/H) \quad \text{for each pair } H \subseteq K,$$

where the above morphism is induced by $B_H = E_G/H \to E_G/K = B_K$.

Examples. As an important example for later development, we shall compute the coefficient system of rational H_G^*-theory for compact Lie groups G.

Reduction 1. Let G be a compact Lie group and G^0 be the identity component of G. Then $\Gamma = G/G^0$ is a *finite* group and

$$B_{G^0} = E_G/G_0 \xrightarrow{p} E_G/G = B_G$$

is a covering space with $\Gamma = G/G^0$ acting as deck transformations.

Hence, it follows from a simple theorem of Grothendieck [G 5] that $p^*: H^*(B_G; \mathbb{Q}) \cong H^*(B_{G^0}; \mathbb{Q})^\Gamma$ (fixed elements under the induced action of Γ).

Lemma (1.1). *Let G be a compact connected Lie group, T be a maximal torus, $N(T)$ be the normalizor of T in G, $W = N(T)/T$ be the Weyl group of G. Then $H^*(G/N(T); \mathbb{Q}) \cong H^*(G/T; \mathbb{Q})^W \cong H^*(pt; \mathbb{Q})$, i.e., $G/N(T) \sim_{\mathbb{Q}} pt$.*

Proof. Since $W \to G/T \to G/N(T)$ is a finite covering, it follows that

$$H^*(G/N(T); \mathbb{Q}) \cong H^*(G/T; \mathbb{Q})^W \quad \text{and} \quad \chi(G/N(T)) = \frac{1}{|W|} \cdot \chi(G/T).$$

On the other hand, the well-known Bruhat decomposition induces a cell-decomposition of G/T with exactly $|W|$ cells of even dimension [p. 347, B 7]. Hence

$$H^{\text{odd}}(G/T; \mathbb{Q}) = 0 \quad \text{and} \quad \dim_{\mathbb{Q}} H^*(G/T; \mathbb{Q}) = \chi(G/T) = |W|$$

and consequently,

$$H^{\text{odd}}(G/N(T); \mathbb{Q}) \cong H^{\text{odd}}(G/T; \mathbb{Q})^W = 0,$$
$$\dim_{\mathbb{Q}} H^*(G/N(T); \mathbb{Q}) = \chi(G/N(T)) = \frac{1}{|W|} \cdot \chi(G/T) = 1, \text{ i.e., } G/N(T) \sim_{\mathbb{Q}} pt. \quad \square$$

Reduction 2. Let G be a compact connected Lie group, T be a maximal torus and W be the Weyl group acting as an automorphism group of T. Then

$$H^*(B_G; \mathbb{Q}) \cong H^*(B_{N(T)}; \mathbb{Q}) \cong H^*(B_T; \mathbb{Q})^W.$$

Proof. Since the fibre of the bundle $G/N(T) \longrightarrow B_{N(T)} \xrightarrow{\pi} B_G$ is $\mathbb{Q}$-acyclic, it follows easily from Serre spectral sequence that $\pi^*: H^*(B_G; \mathbb{Q}) \to H^*(B_{N(T)}; \mathbb{Q})$ is an isomorphism. Hence, one has

$$H^*(B_G; \mathbb{Q}) \cong H^*(B_{N(T)}; \mathbb{Q}) \cong H^*(B_T; \mathbb{Q})^W. \quad \square$$

Example 1. $G = U(n)$, Then $T^n = \{\operatorname{diag}(e^{2\pi i\theta_1}, \ldots, e^{2\pi i\theta_n})\}$ is a maximal torus and W acts on T by permuting θ's; and $H^*(B_{T^n}; \mathbb{Q}) \cong \mathbb{Q}[x_1, \ldots, x_n]$ where $\{x_1, \ldots, x_n\}$ are respectively the transgression of the basis of $H^1(T^n; \mathbb{Q})$ corresponding to $\{\theta_1, \ldots, \theta_n\}$. Hence W acts on $H^*(B_{T^n}; \mathbb{Q})$ as permutations of the x's and

$$H^*(B_G; \mathbb{Q}) \cong \mathbb{Q}[x_1, \ldots, x_n]^W \cong \mathbb{Q}[c_1, c_2, \ldots, c_n]$$

is exactly the ring of symmetric polynomials and the universal Chern classes $c_1, \ldots, c_n$ are respectively the elementary symmetric polynomials.

Example 2. $G = \mathrm{SO}(2n+1)$. Then

$$T^n = \left\{ \begin{pmatrix} \begin{matrix} \cos{}_1 & -\sin\theta_1 \\ \sin{}_1 & \cos\theta_1 \end{matrix} & & \bigcirc \\ & \ddots & \\ \bigcirc & & \begin{matrix} \cos\theta_n & -\sin\theta_n \\ \sin\theta_n & \cos\theta_n \end{matrix} & \\ & & & 1 \end{pmatrix} \right\}$$

is a maximal torus and W acts on T^n by permuting θ's and changing signs. Hence

$$H^*(B_{\mathrm{SO}(2n+1)}; \mathbb{Q}) \cong \mathbb{Q}[x_1, \ldots, x_n]^W \cong \mathbb{Q}[p_1, p_2, \ldots, p_n]$$

is the ring of symmetric polynomials in x_j^2, where the universal Pontrjgin classes $p_1, p_2, \ldots, p_n$ are respectively the elementary symmetric polynomials in x_j^2.

Example 2'. $G = \mathrm{SO}(2n)$. Then W acts on T^n by permuting θ's and changing even number of signs. Hence $e = x_1 \cdot x_2 \cdots x_n$ is also fixed under W and

$$H^*(B_{\mathrm{SO}(2n)}; \mathbb{Q}) \cong \mathbb{Q}[p_1, p_2, \ldots, p_{n-1}, e];\ e^2 = p_n, e \text{ is the universal Euler class.}$$

(C) *Spectral Sequences Related to the Equivariant Cohomology Theory*

In the above construction of Borel, the space X_G is constructed together with two canonical projections, namely, $\pi_1: X_G \to B_G$ and $\pi_2: X_G \to X/G$. Therefore, in the framework of cohomology theory, there are the *Serre spectral sequence* of the fibre map π_1 and the *Leray spectral sequence* of π_2 that offer useful ways to study the equivariant cohomology $H_G^*(X) = H^*(X_G)$.

Serre spectral sequence of π_1. For a given fibration: $X \xrightarrow{i} M \xrightarrow{\pi} B$ over a cell complex B, the skeleton filtration, $\{B^p = p\text{-skeleton of } B\}$, of the base space B lifts to a filtration, $\{M^p = \pi^{-1}(B^p)\}$, of the total space M. Following the usual procedure of constructing a spectral sequence from a space with a given filtration, one obtains the Serre spectral sequence of the fibration which is the main tool for analyzing the cohomological (or homological) relationship between fibre, base and the total space. The following are some of the basic facts which are useful in explicit computations of $H_G^*(X)$. (We refer to [E 3, M 1, S 3] for a thorough discussion of spectral sequences.)

The Serre spectral sequence consists of a sequence of *bigraded differential* algebras $\{(E_n^{p,q}, d_n); n \geq 1\}$ such that

(i) $d_n: E_n^{p,q} \to E_n^{p+n, q-n+1}$ has bigrade $(n, -n+1)$, $d_n^2 = 0$ and the homology of $(E_n^{p,q}, d_n)$ is exactly $(E_{n+1}^{p,q})$.

(ii) $E_1^{p,q} = C^p(B, H^q(X))$ and $E_2^{p,q} = H^p(B, \underline{H^q(X)})$ where $\underline{H^q(X)}$ is the local system of cohomology of fibres.

(ii)′ In the special case of $X_G \xrightarrow{\pi_1} B_G$ with connected G and rational coefficients, then it follows from the simply connectedness of B_G and Kunneth formula that $E_2^{p,q} = H^p(B, \mathbb{Q}) \otimes H^q(X; \mathbb{Q})$.

(iii) $E_n^{p,q} = 0$ for $p < 0$ or $q < 0$, and $E_n^{p,q} = E_{n+1}^{p,q}$ for $n > (p+q+1)$. Hence $E_\infty^{p,q} = E_n^{p,q}$ for $n > (p+q+1)$ is well defined. Moreover, $(E_\infty^{p,q})$ is the associated graded algebra of $H^*(M)$ w.r.t. the filtration

$$F^p H^*(M) = \ker\{H^*(M) \to H^*(M^{p-1})\},$$

namely

$$E_\infty^{p,q} = F^p H^{p+q}(M) / F^{p+1} H^{p+q}(M)$$

(iv) The following two edge homomorphisms are respectively the induced homomorphism i^* and π^*, namely

$$i^*: H^*(M) \to E_\infty^{0,*} \subseteq H^*(X);\ E_\infty^{0,*} = Im(i^*),$$

$$\pi^* H^*(B) \to E^{*,0} \subseteq H^*(M);\ E^{*,0} = Im(\pi^*).$$

Leray spectral sequence of π_2. The Leray spectral sequence [B 10] of a map $\pi: Y \to Z$ is a spectral sequence $\{E_n, d_n\}$, i.e., $E_{n+1} = H(E_n, d_n)$, which begins with $E_2 = H^*(Z; \mathscr{S})$ and converges to $H^*(Y)$, where $\mathscr{S}$ is the *coefficient sheaf* over Z with $H^*(\pi^{-1}(z))$ as its stalk over $z \in Z$. In the case of $\pi_2: X_G \to X/G$, the inverse image $\pi^{-1}(x')$ of $x' = G(x) \in X/G$ can easily be computed as follows:

$$\pi_2^{-1}(x') = E_G \times_G (G/G_x) = (E_G \times_G G)/G_x = E_G/G_x = B_{G_x}.$$

Hence the E_2-term of Leray spectral sequence of π_2 is equal to $H^*(X/G, \mathscr{S})$, where the stalk of $\mathscr{S}$ over x' is $H^*(B_{G_x})$. In general, it is extremely difficult to compute the Leray spectral sequence beyond the E_2-term. However, the above E_2-term provides a precise description of how the coefficient system of H_G^*-theory plays its rôle, and consequently, the above knowledge of E_2-term alone turns out to be quite useful.

Next, let us consider the following problem.

Problem. Let X be a given *G-space* and K be a closed subgroup of G. Then the restriction of G-action to K makes X into a K-space. What is the relationship between $H_G^*(X)$ *and* $H_K^*(X)$?

Clearly, one may take $E_K = E_G$ with the restricted K-action. Then, one has the following commutative diagram of fibrations

$$\begin{array}{ccccc} & & X & \longrightarrow & X \\ & & \downarrow & & \downarrow \\ G/K & \longrightarrow & X_K & \longrightarrow & X_G \\ \downarrow & & \downarrow & & \downarrow \\ G/K & \longrightarrow & B_K & \longrightarrow & B_G . \end{array}$$

Exactly for such a geometric situation, Eilenberg-Moore [E3] constructed a spectral sequence $\{E_n, d_n\}$ such that

$$E_n \Rightarrow H^*(X_K) = H^*{}_K(X),$$
$$E_2^{p,q} = \mathrm{Tor}^{p,q}_{H^*(B_G)}(H^*(B_K), H^*(X_G)).$$

The following are some simple important cases that deserve special attention:

Example 1, $K = \{\mathrm{id}\}$. Then the above spectral sequence reduces to

$$E_2^{p,q} = \mathrm{Tor}^{p,q}_{H^*(B_G)}(H^*(pt), H^*(X_G)),\ E_n \Rightarrow H^*(X).$$

Example 2, Let X, Y be two G-spaces, Then $X \times Y$ is a $(G \times G)$-space and its restriction to the diagonal subgroup $G \xrightarrow{\Delta} (G \times G)$ makes $X \times Y$ into a G-space. Hence, one has a spectral sequence with

$$E_2^{p,q} = \mathrm{Tor}^{p,q}_{H^*(B_G \times B_G)}(H^*(B_G), H^*(X_G \times Y_G))$$

and

$$E_n \Rightarrow H^*_G(X \times Y).$$

This is the Kunneth spectral sequence of H^*_G-theory.

Example 3, $K = G^0$ (the identity component of G); $\Gamma = G/G^0$ is finite. Then it follows easily from the fibration $\Gamma \to X_{G^0} \to X_G$ that

$$H^*_G(X; \mathbb{Q}) \cong H^*_{G^0}(X; \mathbb{Q})^\Gamma.$$

Proposition 1. *Let G be a compact connected Lie group, T be a maximal torus of G, $W = N(T)/T$ be the Weyl group of G and X be a G-space. Then*

(i) $$H^*_G(X; \mathbb{Q}) \cong H^*_{N(T)}(X; \mathbb{Q}) \cong H^*_T(X; \mathbb{Q})^W,$$

(ii) $$H^*_T(X; \mathbb{Q}) \cong H^*_G(X; \mathbb{Q}) \otimes_{H^*_G(pt)} H^*_T(pt) = H^*_G(X; \mathbb{Q}) \otimes_{H^*(B_G, \mathbb{Q})} H^*(B_T, \mathbb{Q}).$$

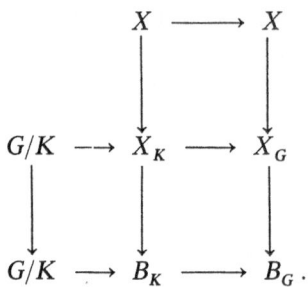

Proof. (i) follows from the $\mathbb{Q}$-acyclicity of $G/N(T)$ and the Serre spectral sequence of the fibration $G/N(T) \to X_{N(T)} \to X_G$.

(ii) follows from the above Eilenberg-Moore spectral sequence and the fact that $H^*(B_T, \mathbb{Q}) \cong H^*(B_G; \mathbb{Q}) \otimes_{\mathbb{Q}} H^*(G/T; \mathbb{Q})$ is a *free-*$H^*(B_G; \mathbb{Q})$ *module.* Thus, the E_2-term reduces to one line;

$$E_2^{p,q} = 0 \quad \text{for} \quad q \neq 0 \quad \text{and}$$

$$E_2^{*,0} = \mathrm{Tor}^{*,0}_{H^*(B_G,\mathbb{Q})}(H^*(B_T; \mathbb{Q}), H_G^*(X; \mathbb{Q})) = H_G^*(X; \mathbb{Q}) \otimes_{H^*(B_G; B)} H^*(B_T; \mathbb{Q}). \quad \square$$

Remark. In view of the maximal torus theorem and the unique central rôle of torus groups in the cohomology theory of transformation groups, the above simple neat result is in fact of basic importance.

§ 2. Localization Theorem of Borel-Atiyah-Segal Type

Let $R = H^*(B_G) = H_G^*(pt)$. Then it is clear that all $H_G^*(X, Y)$ are modules over R and all induced H_G^*-morphisms are morphisms of R-modules. Recall that an element m of an R-module M is said to be (R-)torsion if there exists $r \neq 0 \in R$ such that $r \cdot m = 0$.

Historically, the following simple fact plays a crucial rôle in Borel's approach to the cohomology theory of transformation groups [B 10].

Proposition 2 (A. Borel). *Let G be the circle group, X be a finite dimensional G-space, $F = F(G, X)$ be the fixed point set of X. Then*

(i) $H_G^*(X - F; \mathbb{Q}) = H^*((X - F)/G; \mathbb{Q})$ *is a torsion R-module.*

(ii) *the* Ker *and* Coker *of* $H_G^*(X, \mathbb{Q}) \xrightarrow{r^*} H_G^*(F; \mathbb{Q}) = R \otimes H^*(F; \mathbb{Q})$ *are both torsion R-modules.*

Proof. Since G_x are finite groups for all $x \in (X - F)$, $H^*(B_{G_x}; \mathbb{Q}) = \mathbb{Q}$ for all $x \in (X - F)$. Therefore, the E_2-term of the Leray spectral sequence of $\pi_2 : (X - F)_G \to (X - F)/G$ reduces to one line, namely

$$E_2^{p,q} = 0 \quad \text{if} \quad q \neq 0 \quad \text{and} \quad E_2^{p,0} = H^p((X - F)/G; \mathbb{Q}).$$

Hence $H_G^*(X - F; \mathbb{Q}) \cong H^*((X - F)/G; \mathbb{Q})$ which is finite dimensional and obviously R-torsion. Clearly. (ii) follws directly from (i) and the exact sequence of the pair (X, F) in H_G^*-theory. $\square$

Algebraically, the above proposition can easily be reformulated into the following formally neater statement, namely,

$$H_G^*(X; \mathbb{Q}) \otimes_R \hat{R} \xrightarrow{i^* \otimes \hat{R}} H_G^*(F; \mathbb{Q}) \otimes_R \hat{R} = H^*(F; \mathbb{Q}) \otimes_{\mathbb{Q}} \hat{R}$$

is an isomorphism, where $\hat{R}$ is the quotient field of R. This is the primitive version of localization theorems of Atiyah-Segal type [A 8]. Recall that a multiplicative semigroup, $S \supseteq \{1\}$, contained in the center of R is called a *multiplicative system*; and the localized module $S^{-1}M$ of an R-module M consists of all fractions

$\{m/s, m\in M, s\in S\}$ with the usual identification $m_1/s_1 = m_2/s_2$ if and only if $ss_1 m_2 = ss_2 m_1$ for some $s\in S$. Then $S^{-1}M$ is an $(S^{-1}R)$-module and $M\mapsto S^{-1}M$ is an *exact functor*. Now, let us consider the case $R = H^*(B_G) = H_G^*(pt)$ and $M = H_G^*(X, Y)$. For a given multiplicative system $S\subseteq R$, we set

$$X^S = \{x\in X \mid \text{no element of } S \text{ maps to zero in } R\to H^*(B_{G_x})\}.$$

Theorem (III.1) (Localization). *Let G be a compact Lie group and X be a compact G-space. Then the localized restriction homomorphism*

$$S^{-1}H_G^*(X)\to S^{-1}H_G^*(X^S)$$

is an isomorphism.

Proof. (i) We shall first prove the special case $X^S = \emptyset$. In this case, we need only to show that there exists $s\in S$ such that $\pi_1^*(s) = 0$ in $H_G^*(X)$. By the slice theorem [cf. Th. (I.5), § 3, Ch. I], each orbit $G(x)$ has invariant open neighborhood U such that $G(x)$ is an equivariant retract of U. By compactness of X, there are finite number of such neighbornhoods $\{U_1, \dots, U_q\}$ covering X, with U_i retracting to $G(x_i)$. Since X^S is assumed to be empty, there is an $s_i\in S$ which maps to zero in $H^*(B_{G_{x_i}}) = H_G^*(G(x_i))$ and hence also in $H_G^*(U_i)$. Then it is clear that $\pi_1^*(s_1\cdot \cdots \cdot s_q) = 0$ in $H_G^*(U_1\cup\cdots\cup U_q) = H_G^*(X)$.

(ii) The general case is equivalent to showing $S^{-1}H_G^*(X, X^S) = 0$ which means that for every $x\in H_G^*(X, X^S)$ there is an $s\in S$ with $s\cdot x = 0$. We may assume that

$$x\in H_G^n(X, X^S) = H^n(X_G, X_G^S) = H^n(\pi^{-1}(B_G^k), X_G^S\cap\pi^{-1}(B_G^k) \quad \text{for} \quad k>n.$$

Since $\pi^{-1}(B_G^k)$ is compact and the neighborhood of $X_G^S\cap\pi^{-1}(B_G^k)$ in $\pi^{-1}(B_G^k)$ of the form $\{V_G\cap\pi^{-1}(B_G^k);\ V \text{ invariant neighborhood of } X^S\}$ are cofinal, it follows from the continuity of Čech cohomology that there exists an invariant neighborhood V of X^S such that

$$x\in \operatorname{Im}\{H_G^*(X, V)\to H_G^*(X, X^S)\}.$$

On the other hand, there exist an invariant compact subspace $Y\subseteq (X - X^S)$ such that $V\cup \operatorname{int}(Y) = X$. Hence, it follows from (i) and the fact $Y^S = \emptyset$ that there is $s\in S$ with $\pi_1^*(s)\in \operatorname{Im}\{H_G^*(X, Y)\to H_G^*(X, \operatorname{int} Y)\to H_G^*(X)\}$. Therefore $s\cdot x$ lies in the image of

$$0 = H_G^*(X; V\cup \operatorname{int} Y)\to H_G^*(X, X^S), \quad \text{i.e.,} \quad s\cdot x = 0. \quad \square$$

Theorem (III.1′). *Let G be a compact Lie group and X be a G-space, paracompact and with finite cohomological dimension. Let $S\subseteq H^*(B_G)$ be a multiplicative system and $s\in S$. Then the localized restriction homomorphism*

$$S^{-1}H_G^*(X)\to S^{-1}H_G^*(X^s)$$

is an isomorphism. If X consists of only finite orbit types, then

$$S^{-1} H_G^*(X) \to S^{-1} H_G^*(X^S)$$

is also an isomorphism.

Proof. We shall first prove the case $X^s = \emptyset$. Since X is assumed to be of finite cohomological dimension, it is not difficult to show that the cohomological dimension $\mathrm{cd}(X/G) \leqslant \mathrm{cd}(X)$ is also finite [see Q1]. Hence the E_2-term of Leray spectral sequence of $\pi_2: X_G \to X/G$ is bounded from the right, i.e., $E_2^{p,q} = 0$ for p bigger than a fixed, sufficiently large N. Therefore,

$$E_\infty^{p,q} = 0 \quad \text{for} \quad p > N$$

and there is a decreasing filtration $F^p H_G^*(X)$ with $F^{N+1} H_G^*(X) = 0$ satisfying:

(i) $a \in F^p, b \in F^{p'}$ imply $a \cdot b \in F^{p+p'}$,

(ii) $$E_\infty^{p,*} = F^p H_G^*(X)/F^{p+1} H_G^*(X).$$

The assumption $X^s = \emptyset$ simply means s maps to zero in every stalk $\mathscr{S}_{x'} = H^*(B_{G_x})$. Hence s maps to zero in

$$E_2^{0*} = H^0(X/G; \mathscr{S}) \quad \text{and also in} \quad E_\infty^{0,*}.$$

Therefore, it follows from the following exact sequence

$$0 \to F^1 \to F^0 = H_G^*(X) \to E_\infty^{0,*} \to 0$$

that $\pi_1^*(s) \in F^1 H_G^*(X)$. Then, it follows easily that

$$\pi_1^*(s^{N+1}) \in F^{N+1} H_G^*(X) = 0, \quad \text{i.e.,} \quad \pi_1^*(s^{n+1}) = 0$$

which clearly implies $S^{-1} H_G^*(X) = 0$ (for $s^{N+1} \in S$).

The transition from the case $X^s = \emptyset$ to the case $X^s \neq \emptyset$ is the same as in the compact case. We shall show that

$$S^{-1} H_G^*(X) \to S^{-1} H_G^*(X^S)$$

is also an isomorphism under the assumption of finite orbit types. Let $G_1, G_2, \ldots, G_n$ be the orbit types in $X - X^S$ and $s_i \in S$ maps to zero in $H^*(B_{G_i})$. Then it is clear that

$$X^S = X^s \quad \text{for} \quad s = s_1 \cdot s_2 \cdot \cdots \cdot s_n$$

and hence the above proof applies. □

Remark. (i) The above proof in fact shows that

$$s^{N+1} H_G^*(X, X^s) = 0 \quad \text{for} \quad N \geqslant \mathrm{cd}(X),$$

which is sometimes a useful fact.

(ii) Heuristically, both the statement and the proof of the above localization theorem can be obtained by applying the localization functor S^{-1} to the Leray spectral sequence of π_2. Since the localization functor S^{-1} is exact, it is reasonable to expect that

$$S^{-1}E_2 = H^*(X/G; S^{-1}\mathscr{S}),\ (S^{-1}\mathscr{S})_{x'} = S^{-1}(\mathscr{S}_{x'}) = S^{-1}H^*(B_{G_x})$$

and $\{S^{-1}E_n\} \Rightarrow S^{-1}H_G^*(X)$.

Hence, $S^{-1}\mathscr{S}_{x'} = 0$ for $x \in X - X^S$ and it follows easily that $S^{-1}H_G^*(X, X^S) = 0$.

Chapter IV. The Orbit Structure of a G-Space X and the Ideal Theoretical Invariants of $H_G^*(X)$

In this chapter, we shall proceed to investigate the relationship between the *geometric structures* of a given G-space X and the algebraic structures of its equivariant cohomology $H_G^*(X)$. From the viewpoint of transformation groups, those structures which are usually summarized as the *orbit structure* are certainly the most important geometric structures of a given *G-space*. Hence, it is almost imperative to investigate how much of the orbit structure of a given G-space X can actually be determined from the algebraic structure of its equivariant cohomology $H_G^*(X)$. To be more precise, let us formulate a few more specific problems as examples:

Problem 1. How much of the cohomology structure of the fixed point set, $H^*(F)$, is determined by the equivariant cohomology $H_G^*(X)$?

Problem 2. Is it possible to give a criterion for the *existence* of fixed points purely in terms of the equivariant cohomology $H_G^*(X)$?

Problem 3. Suppose $F(G,X)=\emptyset$. How to determine the set of *maximal* isotropy subgroups, $\{H_i \subseteq G$; maximal among those H with $F(H,X) \neq \emptyset\}$ from the algebraic structure of $H_G^*(X)$?

One of the most profound as well as fascinating facts in the cohomology theory of transformation groups is the following sharp contrast of behaviours between elementary abelian groups, i.e., torus T^k and p-torus $\mathbb{Z}_p^k$, and the rest of compact Lie groups. For example, in the case of torus or p-torus, there are strong regularity theorems which provide clear cut answers to the above problems [H 17]; but for all the other compact Lie groups G, there are wild counter examples which clearly indicate the *non-existence* of a general relationship between the *orbit structure* of X and the *algebraic structure* of $H_G^*(X)$. In retrospect, this also explains why the torus groups play such a central rôle in the representation theory of compact connected Lie groups [cf. Ch. II], which, after all, is concerned with the special case of *linear* transformation groups.

§ 1. Some Basic Fixed Point Theorems

In this section, we consider the algebraic structure of equivariant cohomology, $H_G^*(X)$, as *given* and proceed to investigate that how much of the cohomology ring of the fixed point set, $H^*(F)$, can be determined purely in terms of the equivariant cohomology. First, let us mention the following "counter-examples" of [§1, H16] which clearly indicates the *non-existence* of any general relationship between $H^*(F)$ and $H_G^*(X)$ *unless G is torus or p-groups.*

(A) *An Important Example*

Let G be a compact connected *non-abelian* Lie group and K be an arbitrary finite complex. Then there exists a *compact, finite-dimensional, acyclic G*-space X whose fixed point set $F(G,X)$ is exactly the arbitrarily given complex K. Hence, in particular, its equivariant cohomology $H_G^*(X)$ is the same as that of a single point, i.e.,

$$H_G^*(X) \cong H_G^*(pt) \cong H^*(B_G),$$

but the cohomology ring of its fixed point set $H^*(F) \cong H^*(K)$ is completely arbitrary.

The construction of such a G-space consists of the following steps:

(1) Let $\mathbb{R}^{(2n^2+3n)}$ be the space of $(2n+1)\times(2n+1)$ symmetric matrices of *zero trace* with the usual $SO(2n+1)$-action via conjugation. Let $S^N, N=(2n^2+3n-1)$, be its unit sphere with the restricted $SO(2n+1)$-action. Then the orbit structure of such a G-space is quite simple and an *equivariant* map of *degree zero* $f: S^N \to S^N$ can be explicitly constructed. We refer to p. 717 of [H 5,I] for the explicit computation of orbit structure and construction of such an equivariant map with degree zero.

(2) It is not difficult to prove the following proposition.

Proposition [p. 715, H5, I]. A compact connected Lie group G is *non-abelian* if and only if there exists at least one *odd-dimensional, irreducible*, non-trivial real representation of G.

Let G be a compact, connected, non-abelian Lie group and ψ be an *irreducible* (non-trivial) real representation of dimension $(2n+1)$. Then the restriction of the above $SO(2n+1)$-action on S^N to G (via $G \xrightarrow{\psi} SO(2n+1)$) makes S^N into a G-space *without* fixed point (due to the fact that all connected isotropy subgroups of the above $SO(2n+1)$-action on S^N are *reducible* subgroups).

(3) Based on the above fixed-point-free G-space S^N and the equivariant map $f: S^N \to S^N$ with *degree zero*, set Y to be the *inverse limit* of

$$\cdots \xrightarrow{f} S^N \xrightarrow{f} S^N \xrightarrow{f} S^N \longrightarrow \cdots$$

and $X = Y \circ K$ (the joint of Y and K), equipped with the induced G-action. Then it is easy to verify that X is compact finite-dimensional and acyclic and $F(G,X)=K$.

(B) *A Fundamental Fixed Point Theorem*

In view of the above examples, the *only remaining possibility* of existence of a firm relationship between $H_G^*(X)$ and $H^*(F)$ is the case $\underline{G=T^r \text{ or } \mathbb{Z}_p^r}$. Indeed, in the case $G=T$ or $\mathbb{Z}_p$, we have the following basic result of A. Borel (reformulated in terms of localization):

Let $G=T^r$ (resp. $\mathbb{Z}_p^r$) and $k=\mathbb{Q}$ (resp. $\mathbb{Z}_p$). Then it is well known that

$$\begin{cases} H^*(B_G:k)\cong k[t_1,\dots,t_r],\ \deg t_j=2\ (\text{resp. } 1) \text{ when } G=T \\ \qquad\qquad (\text{resp. } \mathbb{Z}_2^r)\ k=\mathbb{Q}\ (\text{resp. } \mathbb{Z}_2), \\ H^*(B_G,k)\cong k[t_1,\dots,t_r]\otimes\Lambda[v_1,\dots,v_r],\ \deg v_j=1, t_j=\beta\cdot v_j \\ \qquad\qquad \text{when } G=\mathbb{Z}_p^r,\ k=\mathbb{Z}_p \text{ and } p\neq 2. \end{cases}$$

In both cases, we shall denote the polynomial part by R, i.e., $R=k[t_1,\dots,t_r]$ and set $S=R-\{0\}$ which is clearly a multiplicative system lying in the center of $H^*(B_G,k)$.

Proposition 1 (A. Borel). *Let $G=T^r$ or $\mathbb{Z}_p^r, X$ be a paracompact G-space with finite cohomology dimension, and $F=F(G,X)$ be the fixed point set. Then the following localized restriction homomorphism*

$$S^{-1}H_G^*(X,k)\to S^{-1}H_G^*(F,k)=H^*(F,k)\otimes_k(S^{-1}H^*(B_G,k))$$

is an isomorphism.

Proof. Observe that, in the case of $G=T^r$ or $\mathbb{Z}_p^r$,

$$S\cap \operatorname{Ker}\{H^*(B_G,k)\to H^*(B_K,k)\}\neq\emptyset \quad \text{(non-empty)}$$

for any *proper* subgroup $K\subsetneqq G$. Hence, in the terminology of §2, Chapter III, $X^S=F$ and the above proposition is simply a special case of the localization theorem (cf. §2, Ch. III). ▯

As a corollary of the above proposition, we have the following:

Corollary 1. (Criterion for the existence of fixed point): *In the case $G=T^r$ or $\mathbb{Z}_p^r$ and $k=\mathbb{Q}$ or $\mathbb{Z}_p$ respectively, the fixed point set $F(G,X)$ is non-empty if and only if*

$$H_G^*(pt,k)\to H_G^*(X,k)$$

is a monomorphism, i.e., $1\in H_G^*(X,k)$ is "torsion-free".

Proof. Suppose $F\neq\emptyset$ and $q\in F$. Then $B_G\to\{q\}_G\subseteq X_G$ is obviously a cross-section of the fibration $X_G\to B_G$ and consequently

$$H_G^*(pt,k)=H^*(B_G,k)\to H_G^*(X,k)$$

must be a monomorphism. On the other hand, if the above map is a monomorphism, then $1\in H_G^*(X,k)$ is torsion free and hence $S^{-1}H_G^*(X,k)\neq 0$. Therefore, it follows from the above proposition that

$$S^{-1}H_G^*(F,k)\cong S^{-1}H_G^*(X,k)\neq 0$$

which clearly implies $F\neq\emptyset$. ▯

Corollary 2. $\dim_k H^*(F,k) \leqslant \dim_k H^*(X,k)$ *and equality holds if and only if* $E_2 = E_\omega$ *for the Serre spectral sequence.*

Remarks. (i) Let R_0 be the quotient fields of R, i.e., $R_0 = k(t_1, \ldots, t_r)$. Then it is clear that

$$R_0 = S^{-1} R = S^{-1} H^*(B_G, k)/S^{-1} N$$

where N is the set of nilpotent elements of $H^*(B_G, k)$. Hence, the above proposition can also be restated as

$$H_G^*(X) \otimes_{H_G^*(pt)} R_0 \cong H_G^*(F) \otimes_{H_G^*(pt)} R_0 \cong H^*(F) \otimes_k R_0 .$$

(ii) Since all elements of $S = R - \{0\}$ are of even degree (if $p \neq 2$) the localized algebra $S^{-1} H_G^*(X,k)$ or $H_G^*(X) \otimes_{H_G^*(pt)} R_0$ still *preserves* a *mod 2 gradation and* its anti-commutativity if $p \neq 2$.

(iii) Algebraically, $H_G^*(X,k) \otimes_{H_G^*(pt)} R_0$ can simply be considered as the torsion-free part of $H_G^*(X,k)$. Hence, roughly speaking, the above proposition tells us that the torsion-free part of $H_G^*(X,k)$ is geometrically carried by its fixed point set F.

From an algebraic viewpoint, the R_0-algebra $H_G^*(X,k) \otimes_{H_G^*(pt)} R_0$ is *given* in terms of generators and relations. On the other hand, it is more agreeable, from a geometric viewpoint, to obtain the cohomology ring $H^*(F^j, k)$ of each *connected component* F^j of the fixed point set F seperately instead of their total sum $H^*(F,k) \cong \sum H^*(F^j, k)$. To be more precise, let

$$A = R_0[x_1, \ldots, x_l] \otimes_{R_0} \Lambda_{R_0}[v_1, \ldots, v_m]$$

be the free anticommutative R_0-algebra (with mod 2 gradation) generated by even generators $\{x_1, \ldots, x_l\}$ and odd generators $\{v_1, \ldots, v_m\}$. Then, a presentation of the R_0-algebra $H_G^*(X,k) \otimes_{H_G^*(pt)} R_0$ in terms of generators and relations consists of the following epimorphism and its kernel I, namely,

$$I = \operatorname{Ker}(\rho) \subseteq A \cong R_0[x_1, \ldots, x_l] \otimes_{R_0} \Lambda_{R_0}[v_1, \ldots, v_m] \xrightarrow{\rho} H_G^*(X,k) \otimes_{H_G^*(pt)} R_0$$

where ρ maps x's to the even generators and v's to the odd generators respectively. Let us consider the following commutative diagram:

$$\begin{array}{ccc}
A = R_0[x_1, \ldots, x_l] \otimes_{R_0} \Lambda_{R_0}[v_1, \ldots, v_m] & \xrightarrow{\rho} & H_G^*(X,k) \otimes_{H_G^*(pt)} R_0 \cong \oplus \sum_{j=1}^{s} H^*(F^j, k) \otimes_k R_0 \\
\downarrow \pi & \searrow^{\rho_j} & \downarrow \rho_j \\
R_0[x_1, \ldots, x_l] & & H^*(F^j, k) \otimes_k R_0
\end{array}$$

Let $I = \operatorname{Ker}(\rho)$, $I_j = \operatorname{Ker}(\rho_j)$, $\bar{I} = \pi(I)$, $\bar{I}_j = \pi(I_j)$. Then, our problem is how to compute I_j from I. We state the answer as follows:

Theorem (IV.1) [H17]. *Let* $\{\xi_1,\ldots,\xi_l;\nu_1,\ldots,\nu_m\}$ *be a generator system of the* R_0*-algebra* $H_G^*(X,k)\otimes_{H_G^*(pt)}R_0$, *and* I *be the ideal of defining relations, namely* $I=\mathrm{Ker}(\rho)$ *of the following epimorphism*:

$$\rho: A=R_0[x_1,\ldots,x_l]\otimes_{R_0}\Lambda_{R_0}[v_1,\ldots,v_m]\to H_G^*(X,k)\otimes_{H_G^*(pt)}R_0.$$

Then (i) *the radical of* I, $\sqrt{I}$, *decomposes into the intersection of* s *maximal ideals* $M_j=M(\alpha_j)$ *whose varieties are respectively the rational points* $\alpha_j=(\alpha_{j1},\alpha_{j2},\ldots,\alpha_{jl})\in R_0^l$, *i.e.*,

$$\sqrt{I}=M_1\cap\cdots\cap M_s,\qquad V(I)=\{\alpha_1,\ldots,\alpha_s\}\subseteq R_0^l\,.$$

(ii) *There is a one-to-one correspondence between the connected component of the fixed point set* $F=F^1+\cdots+F^s$ *and the above points* $\{\alpha_1,\ldots,\alpha_s\}$ *such that the restriction homomorphism of an arbitrary point* $q_j\in F^j\subseteq X$ *maps* $\xi_i\in H_G^*(X,k)$ *to* $\alpha_{ji}\in H_G^*(q_j,k)$.

(iii) $H^*(F^j,k)\otimes_k R_0\cong A/I_j$ *where* $I_j=I_{M_j}\cap A$, I_{M_j} *is the localization of* I *at* M_j,

(iv) $I=I_1\cap\cdots\cap I_s=I_1\cdot I_2\cdots I_s$.

Proof. Let $I=\mathrm{Ker}(\rho)$, $I_j=\mathrm{Kerb}(\rho_j)$. Then the isomorphism

$$H_G^*(X,k)\otimes_{H_G^*(pt)}R_0\xrightarrow{\cong}\oplus\textstyle\sum_{j=1}^s H^*(F^j,k)\otimes_k R_0$$

implies that

$$I=I_1\cap\cdots\cap I_s$$

and

$$I_j+I_1\cap\cdots\cap I_{j-1}\cap I_{j+1}\cap\cdots\cap I_s=1\,.$$

Hence, it follows easily that $I=I_1\cdot I_2\cdot\cdots\cdot I_s$.

Surely, one may assume that $\xi_1,\ldots,\xi_l;\nu_1,\ldots,\nu_m$ are actually elements of $H_G^*(X,k)$. Let q_j be an arbitrary fixed point which lies in the j-th component F^j and let ι_j^* be the restriction homomorphism induced by the inclusion of $q_j\in F^j\subseteq X$. Set $\iota_j^*(\xi_i)=\alpha_{ji}$. Then

$$\rho_j(x_i-\alpha_{ji})\in\tilde{H}^*(F^j,k)\otimes_k R_0$$

and it follows from the fact that every element of the reduced cohomology ring $\tilde{H}^*(F^j,k)$ is nilpotent, $(x_i-\alpha_{ji})^N\in I_j$ for sufficiently large N. In other words, $\sqrt{I_j}=M(\alpha_j)=$ the ideal generated by $\{(x_i-\alpha_{ji})\}$ and $\{v_i\}$, which is clearly maximal. Furthermore, it follows from

$$I_j+I_1\cap\cdots\cap I_{j-1}\cap I_{j+1}\cap\cdots\cap I_s=1$$

that $M_1,\ldots,M_s$ are mutually distinct. Hence, one has $V(I)=\{\alpha_1,\ldots,\alpha_s\}$ and $\sqrt{I}=\sqrt{I_1}\cap\cdots\cap\sqrt{I_s}=M_1\cap\cdots\cap M_s$.

Since $(A-M_j)$ contains no zero divisors of A, it is clear that A can be considered as a subset of A_{M_j}. We claim that

$$I_j=I_{M_j}\cap A\,.$$

Since it is obvious that $(I_h)_{M_j}=1$ for $h\neq j$, one has

$$I_{M_j}=(I_1)_{M_j}\cap\cdots\cap(I_s)_{M_j}=(I_j)_{M_j}$$

and

$$I_j\subseteq(I_j)_{M_j}\cap A=I_{M_j}\cap A.$$

Next let us show the converse inclusion $I_j\supseteq I_{M_j}\cap A$. Notice that there exists a sufficiently large N such that $M_j^N\subseteq I_j$ (for A is Noetherian). Let a be an arbitrary element of $I_{M_j}\cap A$. Then, by definition of localization, there exists $b\in(A-M_j)$ such that $b\cdot a\in I_j$. Since $A/M_j\cong R_0$ is a field, we may assume without loss of generality that

$$b=1-b_0+b_1,\quad b_0,b_1\in M_j$$

where b_0 is even and b_1 is odd. Then

$$\begin{aligned}&(1+b_0+\cdots+b_0^{N-1})^2(1-b_0-b_1)(1-b_0+b_1)\cdot a\\&=(1+b_0+\cdots+b_0^{N-1})^2(1-b_0)^2\cdot a=(1-b_0^N)^2\cdot a\\&=a+b_0^{2N}\cdot a-2b_0^N\cdot a\in I_j.\end{aligned}$$

But $b_0^N\in M_j^N\subseteq I_j$, we have $a\in I_j$, i.e., $I_{M_j}\cap A\subseteq I_j$. Hence $I_j=I_{Mj}\cap A$ and the proof of the theorem is complete. □

(C) *Some Elementary Operations*

The above theorem exhibits a clear cut method of computing a presentation of the R_0-algebra $H^*(F^j,k)\otimes_k R_0$ in terms of generators and relations for each connected component, F^j, of the fixed point set F. However, what we actually want is a presentation of the k-algebra $H^*(F^j,k)$ itself in terms of generators and relations. A simple-minded approach to deal with such an algebraic problem is to try to transform a given system of generators $\{\eta_i\}$ of the R_0-algebra $H^*(F^j,k)\otimes_k R_0$ by some "elementary operations" so that one obtains, eventually, a new system of generators $\{\tilde\eta_i\}$ (of the R_0-algebra) which completely lies in $H^*(F^j,k)$. Then it is clear that $\{\tilde\eta_i\}$ also constitutes a generator system of $H^*(F^j,k)$. We shall show that the following two types of *elementary operations* suffice:

(i) To multiply a generator by an invertible element, i.e., $\eta_i\mapsto\eta_i'=a\eta_i$ where a is invertible.

(ii) To replace a generator, say η_1, by $\eta_1'=\eta_1+f(\eta_2,\ldots,\eta_l)$, where $f(\eta_2,\ldots,\eta_l)$ is a polynomial of $\eta_2,\ldots,\eta_l$.

Proposition 2. *Let $\{\eta_1,\ldots,\eta_l\}$ be a generator system of the R_0-algebra $H^*(F^j,k)\otimes_k R_0$. Then there exist a series of suitable elementary operations of the above two types which transform $\{\eta_i\}$ into a new generator system $\{\tilde\eta_i\}$ which completely lies in $H^*(F^j,k)$.*

Proof. Since the gradation of $H^*(F^j,k)$ and the nilpotency of $\tilde{H}^*(F^j,k)$ play an important rôle, we shall express

$$\eta_i \in \tilde{H}^*(F^j,k) \otimes_k R_0 \cong \oplus \sum_{d>0} H^d(F^j,k) \otimes_k R_0$$

in ascending degrees as follows:

$$\eta_i = \sum \lambda_{i,k} \otimes a_{i,k}; \deg \lambda_{i,k} < \deg \lambda_{i,k+1} .$$

Moreover, we may assume that

$$d_1 = \deg \lambda_{1,1} = \cdots = \deg \lambda_{i_1,1} < \deg \lambda_{i_1+1,1} \leqslant \cdots \leqslant \deg \lambda_{l,1}$$

and $\{\lambda_{1,1}, \ldots, \lambda_{i_1,1}\}$ are linearly independent in $H^{d_1}(F^j,k)$, for otherwise, we can simply apply operation of the second type to get another generator system with smaller i_1. Express $\lambda_{1,1}$ in terms of the generator system $\{\eta_i\}$, and it is not difficult to see that

$$\lambda_{1,1} = \eta_1(a + g_1(\eta)) + f_1(\eta_2, \ldots, \eta_l)$$

where $a \in R_0$ and $g_1(\eta)$ is a polynomial of η *without* constant term. Notice that $g_1(\eta)$ is nilpotent and hence $(a+g_1(\eta))$ is invertible. Therefore, one may first change η_1 to $\eta_1' = \eta_1(a+g_1(\eta))$ and then to $\lambda_{1,1} = \tilde{\eta}_1 = \eta_1' + f_1(\eta_2, \ldots, \eta_l)$.

Similarly, we transform $\eta_2, \ldots, \eta_{i_1}$ to $\tilde{\eta}_2 = \lambda_{2,1}, \ldots, \tilde{\eta}_{i_1} = \lambda_{i_1,1}$. Then, we shall use $\lambda_{1,1}, \ldots, \lambda_{i_1,1}$ to chop off the lower degree terms of $\eta_i, i > i_1$ as far as possible by operation of the second type. Hence, we may again assume that

$$\eta_i' = \sum \lambda_{i,k}' \otimes a_{i,k}, \quad i = i_1 + 1, \ldots, l;$$

$$d_1 < d_2 = \deg \lambda_{i_1+1,1}' = \cdots = \deg \lambda_{i_1+i_2,1}' < \cdots \leq \deg \lambda_{k,1}' \quad \text{and}$$

$\{\lambda_{i_1+1,1}', \ldots, \lambda_{i_1+i_2,1}'\}$ linearly independent modulo $\langle \lambda_{1,1}, \ldots, \lambda_{i_1,1} \rangle$. Then the same procedure will transform η_{i_1+1} to $\tilde{\eta}_{i_1+1} = \lambda_{i_1+1,1}' \cdots$ and $\eta_{i_1+i_2}$ to $\tilde{\eta}_{i_1+i_2} = \lambda_{i_1+i_2,1}'$. Keep going, and it clear that one will eventually obtain a desired system $\{\tilde{\eta}_i\}$ completely lying in $\tilde{H}^*(F^j,k)$. □

Corollary. *In the case $G = T^r$ or $\mathbb{Z}_p^r$ and $k = \mathbb{Q}$ or $\mathbb{Z}_p$ respectively, if $H_G^*(X,k) \otimes_{H_G^*(pt)} R_0$ is generated by ℓ elements as an R_0-algebras, then the cohomology ring of each connected component of the fixed point set, $H^*(F^j,k)$, is also generated by at most ℓ elements as a k-algebra.*

Proof. It follows immediately from Theorem (IV.1) and the above proposition.

(D) *Examples and Preliminary Applications*

In the following discussion, we shall always assume that the group $G = T^r$ (resp. $\mathbb{Z}_p^r$) and the coefficient ring $k = \mathbb{Q}$ (resp. $\mathbb{Z}_p$). Moreover, we shall adopt the notations: $X \sim_k Y$ to mean $H^*(X,k) \cong H^*(Y,k)$; $X \approx_k Y$ to mean that X and Y are of the same k-homotopy type.

Example 1. The simplest test spaces are of course those *acyclic spaces* $X\sim_k\{pt\}$. In this case, it follows immediately from the Serre spectral sequence that $H_G^*(X,k)\cong H_G^*(pt,k)$. Hence, it follows easily from Theorem (IV.1) that $F(G,X)\sim_k pt$.

Example 2. Next let us consider the case $X\sim_k S^n$:

The E_2-term of Serre spectral sequence of $X_G\to B_G$ consists of two lines, i.e., $E_2^{*,q}=H^*(B_G;k)\otimes H^q(X,k)=0$ if $q\neq 0,n$. Hence the generator $x\in H^n(X,k)$ is transgressive and $E_{n+2}=E_\infty$. Since the kernel of $H^*(B_G,k)\to H^*(X_G,k)$ is exactly the ideal generated by the transgression of x, $d_{n+1}(x)$, it follows from the criterion that $F\neq\emptyset$ if and only if $d_{n+1}(x)=0$; or equivalently, if and only if $E_2=E_\infty$. Hence, in the case $F\neq\emptyset$, $H_G^*(X,k)=H^*(B_G,k)\otimes_k H^*(X,k)$ as a $H^*(B_G,k)$-module, and consequently,

$$H_G^*(X,k)\otimes_{H_G^*(pt)}R_0\cong R_0[x]/(f(x))$$

where $f(x)=x^2$ when n is odd and $p\neq 2; f(x)=x^2+ax+b$ otherwise. Again it follows easily from Theorem (IV.1) that (x^2+ax+b) can be factorized into $(x-\alpha_1)(x-\alpha_2)$ in $R_0[x]$, and the cohomology of F is as follows:

$$F\sim_k S^0 \quad \text{(two points) if} \quad \alpha_1\neq\alpha_2,$$

$$F\sim_k S^r(r>0), \quad \text{otherwise.}$$

This is exactly the famous fixed point theorem of P.A. Smith.

Theorem (IV.2) (Smith). *If* $X\sim_k S^n$, *then* $F\sim_k S^r$ *and* $n\equiv r \bmod 2$ *when* $p\neq 2$.

Example 3. $X\sim_k CP^n, p\neq 2$.

That is $H^*(X,k)\cong k[\xi_0]/(\xi_0^{n+1})$. Notice that $\beta_p\cdot\xi_0=0$ (β_p is the mod p Bockstein operator) and β_p commutes with the differentials of the Serre spectral sequence. It is not difficult to show that $E_2=E_\infty$ for the fibration $X\to X_G\to B_G$, $X\sim_k CP^n$. Therefore

$$H_G^*(X,k)\cong H^*(B_G,k)\otimes_k H^*(X,k)$$

as $H^*(B_G,k)$-modules and $\{1,\xi,\xi^2,\dots,\xi^n\}$ forms a module basis of $H_G^*(X,k)$ where ξ is an arbitrary *lifting* of ξ_0. Hence

$$H_G^*(X,k)\cong H^*(B_G,k)[\xi]/(f(\xi)) \quad \text{(as an } H^*(B_G,k)\text{-algebra)}$$

where $f(\xi)=(\xi^{n+1}+c_1\xi^n+c_2\xi^{n-1}+\dots+c_{n+1})\in H^*(B_G,k)[\xi]$, and consequently

$$H_G^*(X,k)\otimes_{H_G^*(pt)}R_0\cong R_0[\xi]/(f(\xi)).$$

Theorem (IV.1) asserts that $f(\xi)$ can be factorized into the product of linear factors *in* $R_0[\xi]$, however, with the help of Gauss lemma, $f(\xi)$ already can be factorized into linear factors in $H^*(B_G,k)[\xi]$, namely

$$f(\xi)=(\xi-\alpha_1)^{m_1}\cdots(\xi-\alpha_s)^{m_s}, \quad \alpha_j\in H^2(B_G,k);$$

and correspondingly, F consists of s connected components

$$F = F^1 + \cdots + F^s \quad \text{with} \quad F^j \sim_k CP^{(m_j - 1)}.$$

We state the above results as follows which is, in fact, the prototype that leads to the formulation of Theorem (IV.1).

Theorem (IV.3) [H 15]. *Let $G = T^r$ (resp. $\mathbb{Z}_p^r, p \neq 2$), $k = \mathbb{Q}$ (resp. $\mathbb{Z}_p$) and X be a G-space with $H^*(X,k) \cong k[\xi_0]/(\xi_0^{n+1})$. Then*

$$H_G^*(X,k) \cong H_G^*(pt,k)[\xi]/(f(\xi)) \quad (\text{as } H_G^*(pt,k)\text{-algebra})$$

where ξ is an arbitrary, fixed lifting of ξ_0,

$$f(\xi) = (\xi - \alpha_1)^{m_1} \cdots (\xi - \alpha_s)^{m_s}$$

and correspondingly, the fixed point set $F \sim_k CP^{(m_1 - 1)} + \cdots + CP^{(m_s - 1)}$.

Remarks. (i) In case $X = CP^n$ and the G-action is the induced action of a complex linear action φ on $\mathbb{C}^{n+1}$, then one can choose a suitable lifting ξ such that the *weight system* (resp. p-weight system) of φ consists of exactly $\{\alpha_i$ with multiplicity $m_i; i = 1, \ldots, s\}$.

(ii) As we shall see in later discussion (cf. § 1, Ch. VI.), the above $\{\alpha_i, m_i\}$ can be also considered as the topological weight system of the given topological action on X. It not only tells us what the fixed point set looks like but also the whole story of the cohomological behavior of the given G-action on X.

(iii) It is interesting to note that the above result fully demostrates the crucial rôle of multiplicative structure of $H_G^*(X,k)$, since the additive structure of $H_G^*(X,k)$ can not even distinguish a *non-trivial* G-action from the trivial G-action on X.

Example 4. $X \sim_k S^m \times S^n$; m, n even and $p \neq 2$.

In this case, the Serre spectral sequence again has no *non-zero* differentials simply for dimensional reasons. Hence $E_2 = E_\infty$, and

$$H_G^*(X,k) \cong H^*(B_G,k) \otimes_k H^*(X,k)$$

as $H^*(B_G,k)$-modules and $\{1, x, y, x \cdot y \deg x = m, \deg y = n\}$ forms a module basis of $H_G^*(X,k)$. Therefore, multiplicatively

$$H_G^*(X,k) \cong H^*(B_G,k)[x,y]/(f(x,y), g(x,y))$$

where $f(x,y)$ and $g(x,y)$ are equations of parabola type. Theorem (IV.1) asserts that the two parabolas $f(x,y) = 0$ and $g(x,y) = 0$ in R_0^2 intersect at "*rational* points". Hence, geometrically speaking, there are the following possibilities:

(i) two intersecting double lines,

(ii) a double line intersecting with a parabola or parallel line,

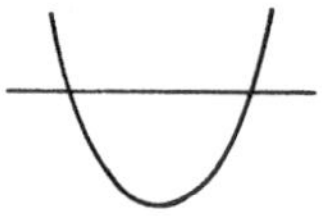

or

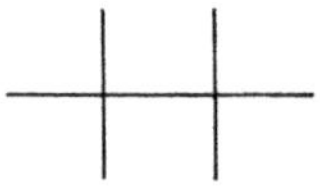

(iii) a double line tangent with a parabola,

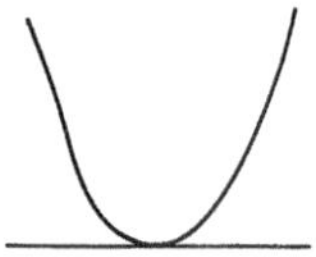

(iv) two parabolas intersect at four points,

or

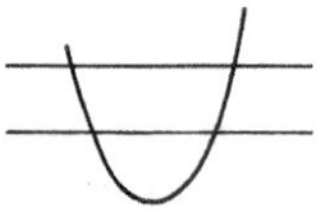

or

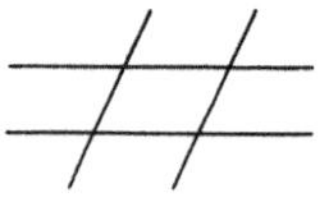

(v) two parabolas intersect at two points and tangent at one point,

or

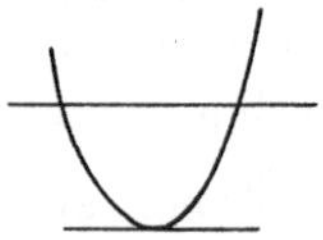

(vi) two parabola intersect at one point and tangent at one point.

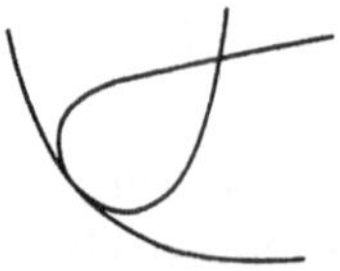

Algebraically, a straight-forward computation according to Theorem (IV.1) will show that the cohomology of the fixed point set are respectively as follows:

(i) $F \sim_k$ product of two even spheres.
(ii) $F \sim_k$ sum of two even spheres (in general of different dimensions).
(iii) $H^*(F,k) \cong k[\xi]/(\xi^4)$, $\deg(\xi)$ even
(iv) $F \sim_k$ four points (or $S^0 \times S^0$).

(v) $F \sim_k$ two points and an even sphere (or S^0 and an even sphere).
(vi) $H^*(F,k) \cong k + k[\xi]/(\xi^3)$.

Theorem (IV.4) [H 17]. *Let $G = T^r$ (resp. $\mathbb{Z}_p^r$, $p \neq 2$), $k = \mathbb{Q}$ (resp. $\mathbb{Z}_p$) and X be a G-space with $X \sim_k S^m \times S^n$; m, n even. Then the possibility of cohomology algebra of the fixed point set F are as follows:*

(i) *$F \sim_k$ product of two even spheres (including $S^0 \times S^0$),*
(ii) *$F \sim_k$ sum of two even spheres (in general of different dimensions),*
(iii) *$H^*(F,k) \cong k[\xi]/(\xi^4)$, $\deg(\xi)$ even,*
(iv) *$H^*(F,k) \cong k \oplus k[\xi]/(\xi^3)$, $\deg(\xi)$ even.*

Remarks. (i) In the case $G = \mathbb{Z}_2^r$, the Serre spectral sequence may have non-zero differentials. However, if one assumes $E_2 = E_\infty$, then the above result holds except for the evenness of degree. We refer to a paper of J. C. Su [S 13] for a thorough discussion of this case.

(ii) In a recent paper of Tomter [T 1], he constructs examples realizing all the above four possibilities.

We conclude this section by the following useful corollary of Theorem (IV.1).

Theorem (IV.5) [S 4]. *Let $G = T^r$ and X be a G-space with vanishing even-dimensional rational homotopy groups. Then the fixed point set F (if non-empty) is always connected.*

Proof. Let q_1, q_2 be two arbitrary fixed points and $\xi \in H_G^*(X, \mathbb{Q})$ be an *arbitrary* element. We shall show that the restriction of ξ to $H_G^*(q_1, \mathbb{Q})$ and $H_G^*(q_2, \mathbb{Q})$ must be the *same* and then it follows from Theorem (IV.1) that q_1, q_2 belong to the same connected component of F, consequently, F is connected (for q_1, q_2 are arbitrary). For technical simplicity, we may reduce the proof to the special case $G = T^1 =$ circle group, because there always exists a suitable circle subgroup $S^1 \subseteq T^r$ with $F(S^1, X) = F(T^r, X)$. Suppose $\deg(\xi) = n$, then it suffices to show that the following two cross-sections σ_1, σ_2 are in a sense "$\mathbb{Q}$-homotopic":

$$\begin{array}{ccccc} \{q_1\}_G^n & \subseteq & X_G^n & \supseteq & \{q_2\}_G^n \\ & \nwarrow \sigma_1 & \downarrow & \sigma_2 \nearrow & \\ & & B_G^n = CP^n & & \end{array}$$

Since CP^n consists of only even cells and X has only torsion even homotopy groups, a straight-forward obstruction theory argument will show that there exists a suitable N such that $\rho_N^{-1}(\sigma_1)$ and $\rho_N^{-1}(\sigma_2)$ are actually homotopic where

$$\rho_N : B_G^n \to B_G^n$$

is induced by the homomorphism $g \mapsto g^N$. Hence

$$\rho_N^* \sigma_1^*(\xi) = \rho_N^* \sigma_2^*(\xi)$$

which obvious implies $\sigma_1^*(\xi) = \sigma_2^*(\xi)$ because ρ_N^* is obviously monomorphic. □

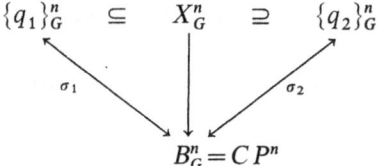

§ 2. Torsions of Equivariant Cohomology and F-Varieties of G-Spaces

In this section, we shall assume, once for all, that $G=T^r$ (resp. $\mathbb{Z}_p^r$) and the coefficient ring $k=\mathbb{Q}$ (resp. $\mathbb{Z}_p$), because this is the only setting where some kind of close relationship between the algebraic invariants of equivariant cohomology and the orbit structures of G-spaces is at all possible. In the last section, we succeed in establishing a clear-cut, definite relationship between the torsion-free part of $H_G^*(X,k)$ and $H^*(F,k)$. Hence, a reasonable rough guess would be that the torsion part of $H_G^*(X,k)$ should be related to orbit structures other than $F(G,X)$, such as the fixed point set $F(K,X)$ of subgroups $K \subseteq G$, etc. To be more precise, let us introduce the following specific notions about the orbit structures.

(A) *Orbit Types and F-Varieties*

For a given G-space X, the orbit type of $G(x)$ is determined by the conjugacy class of its isotropy subgroup G_x. Hence, it is convenient to define the orbit types of a G-space X as follows:

Definition. $\mathcal{O}(X)=$ the conjugacy classes of subgroups in $\{G_x; x\in X\}$, $\mathcal{O}^0(X)=$ the conjugacy classes of subgroups in $\{G_x^0; x\in X\}$, and we shall call $\mathcal{O}(X)$ the orbit type of X, $\mathcal{O}^0(X)$ the *connected* orbit type of X.

Remark. In the special case that G is *abelian*, then each conjugacy class of subgroups consists of a single subgroup and hence $\mathcal{O}(X)$, $\mathcal{O}^0(X)$ are simply the set of isotropy subgroups and connected isotropy subgroups respectively.

The situation of G-spaces is quite analogous to that of algebraic varieties. Hence, one may also define an analog of Zariski closure for G-spaces as follows.

Definition. Let x be a point in a given G-space X. The F-variety of x, denoted by $F(x)$, is defined to be the following invariant subspace:

$$F(x)=\text{the connected component of } x \text{ in } X^{G_x}=\bigcup_K \{F(K,X);\, K \text{ conjugate to } G_x\}.$$

In case G is a compact connected Lie group, it is also useful to have slight variant F^0-variety, $F^0(x)$, as follows

$$F^0(x)=\text{the connected component of } x \text{ in } X^{G_x^0}=\bigcup_K \{F(K,X);\, K \text{ conjugate to } G_x^0\}.$$

Remarks. (i) In the important special case $G=T^r$ or $\mathbb{Z}_p^r$, the isotropy subgroups G_x are automatically normal. Hence, $F(x)$ or $F^0(x)$ are simply the connected component of x in $F(G_x,X)$ or $F(G_x^0,X)$ respectively.

(ii) Intuitively speaking, $X^{G_x}=\bigcup_K \{F(K,X);\, K\sim G_x\}$ is simply the set of all points of higher or equal singularity as that of x. The reason for defining $F(x)$ to be the connected component of x in X^{G_x} rather than the whole set X^{G_x} is based on the belief of topologists that the "closure" of a single point should be connected.

(iii) For a given G-space X, the family of F-varieties of X is the collection of distinguished invariant spaces which are *canonically* defined. In the cohomology theory of transformation groups, the general term of "orbit structure" can be interpreted as the cohomological structure of the "network of F-varieties."

(iv) Let $Y = F(x_0)$ be a given F-variety of X. Then $y \in Y$ is called a *generic point* of Y if and only if $G_y \sim$(conjugate to) G_{x_0} and the isotropy subgroups of generic points are called the *generic isotropy subgroups* of Y. In the special case that G is abelian, the generic isotropy subgroups of Y consists of a single subgroup and it is convenient to denote it simply by G_Y. Analogous to the situation of algebraic geometry, one may also introduce the terminology that a G-space X is *irreducible* if X is itself the F-variety of a suitable generic point x_0, i.e., $X = F(x_0)$.

Examples. (1) In case X is a cohomology manifold over $\mathbb{Q}$ (resp. $\mathbb{Z}_p$) and $G = T^r$ (resp. $\mathbb{Z}_p^r$), then every F^0-variety Y (resp. F-variety) of X is also a connected cohomology manifold over $\mathbb{Q}$ (resp. $\mathbb{Z}_p$) and X itself is irreducible if and only if X is connected. [The later assertion is equivalent to the principal orbit type theorem of Montgomery-Samelson-Yang (cf. Theorem (1.7), §3-D, Ch. I].

(2) Suppose $X = \mathbb{R}^n$, $G = T^r$ and the G-action is given by a linear representation φ. Let the system of non-zero weights of φ be

$$\Omega_0(\varphi) = \{\pm w_i; m_i\},$$

(for weights of real representation are in pairs) and $r = \dim F(G, X)$ be the multiplicity of the zero weight. Then

(i) A subgroup $K \in \mathcal{O}^0(X)$ if and only if K is of the form

$$K = w_{i_1}^{\perp} \cap w_{i_2}^{\perp} \cap \cdots \cap w_{i_s}^{\perp} \qquad (K = G \text{ if } s = 0)$$

where $w_i^{\perp} = \mathrm{Ker}(w_i)$ is the corank one subtorus whose Lie algebra is exactly the hyperplane perpendicular to w_i.

(ii) The F^0-variety of x, $F^0(x)$ is a linear subspace whose dimension is given by

$$\dim F^0(x) = r + \sum_{w_i^{\perp} \supseteq G_x^0} 2m_i .$$

(2′) Suppose $X' = S^{n-1}$ is the unit sphere of $\mathbb{R}^n$ in the above example. Then, restriction of the above F^0-variety structure to S^{n-1} gives the F^0-variety structure in S^{n-1}, namely, $F^0(x)$ is a sphere of dimension $(r + \sum_{w_i^{\perp} \supseteq G_x^0} 2m_i - 1)$.

(3) Suppose $X = CP^n$, $G = T^r$ and the G-action is induced from the complex linear representation φ with weight system $\Omega(\varphi) = \{w_i, m_i\}$. Then:

(i) The fixed point set $F(G, X) = CP^{(m_1 - 1)} + \cdots + CP^{(m_s - 1)}$ which are indexed by the different weights $w_1, \ldots, w_s$ respectively.

(ii) Let $F_{w_{i_1}}, \ldots, F_{w_{i_a}}$ be those connected components of $F(G, X)$ which are included in $F^0(x)$. Then the Lie algebra of G_x^0 is given by

$$w_{i_1} = w_{i_2} = \cdots = w_{i_a},$$

and $F^0(x)$ is a CP^k with $k = (m_{i_1} + m_{i_2} + \cdots + m_{i_a}) - 1$.

(4) Suppose X has no even rational homotopies and $G=T^r$. Then the F^0-varieties of X are bijectively indexed by $\mathcal{O}^0(X)$ [Theorem (IV.5)].

(B) *An Important Testing Case*

In §1, we know that the *torsion-free* part of $H_G^*(X,k)$ is geometrically carried by the fixed point set F. Hence, a natural simplification in the preliminary investigation of the *torsion part* will be to suppress the *torsion-free part* entirely by assuming $F=\phi$ (empty set). Then, it follows from Corollary 1 of §1 that the kernel of $H_G^*(pt,k) \xrightarrow{\pi_1^*} H_G^*(X,k)$ is non-zero, which can also be considered as the annahilator of $1 \in H_G^*(X,k)$. The geometric meaning of the above ideal $J=\mathrm{Ker}(\pi_1^*) \subseteq H^*(B_G,k)$ is given by the following theorem of [H17] which is a prototype of further results concerning torsions of equivariant cohomology.

Theorem (IV.6). *Suppose* $G=T^r$ *(resp.* $\mathbb{Z}_p^r$*) and* X *is a compact G-space without fixed point. Let* J *be the kernel of* $\pi_1^*: H(B_G,k) \to H_G^*(X,k)$, $\sqrt{J}$ *be the radical of* J *and*

$$\sqrt{J}=P_1 \cap \cdots \cap P_a$$

be the irreducible decomposition of $\sqrt{J}$ *into its prime components. Then*

(i) *There is a natural bijection between the above prime components,* $\{P_j\}$, *and the maximal elements,* $\{H_j\}$, *of* $\mathcal{O}^0(X)$ *(resp.* $\mathcal{O}(X)$*) such that the variety of* P_j, $V(P_j)$, *is exactly the Lie algebra of* H_j.

(ii) *Let* $Y^j=F(H_j,X); j=1,\ldots,a$. *Then*

$$H_G^*(X,k)_{P_j} \cong H_G^*(Y^j,k)_{P_j} \cong H^*(Y^j/G,k) \otimes_k H^*(B_{H_j},k)_{P_j}.$$

Remark. Combining (ii) with Theorem (IV.1) will give us a powerful grip on the structure of $H^*(Y^j/G,k)$.

Proof. Since the proof for the case $G=\mathbb{Z}_p^r$ is essentially the same as that of the case $G=T^r$ except for some simple modifications, we shall only give here the proof of the case $G=T^r$.

(i) For a given point $x \in X$, let P_x be the kernel of $H^*(B_G,k) \to H^*(B_{G_x},k)$ which is obviously a prime ideal with the Lie algebra of G_x as its variety. For a given prime ideal $P \subseteq H^*(B_G,k)$, let

$$\{X^P = x \in X \mid P_x \subseteq P\}.$$

Notice that, under the localization at $P, H^*(B_{G_x},k)_P=0$ for all $x \notin X^P$. Hence $H_G^*(X,k)_P \cong H_G^*(X^P,k)_P$.

Observe that $P_1,\ldots,P_a$ are exactly those *minimal* prime ideals containing J, and $J \subseteq P_j$ clearly implies that $J_{P_j} \subsetneqq H^*(B_G,k)_{P_j}$. Hence the localized morphism

$$H^*(B_G,k)_{P_j} \to H_G^*(X,k)_{P_j} \cong H_G^*(X^{P_j},k)_{P_j}$$

is non-trivial, and consequently $X^{P_j} \neq \emptyset$ (non-empty). Let $x_j \in X^{P_j}$ and $H_j=G_{x_j}^0$. Then one has the following commutative diagram of exact sequences:

$$\begin{array}{ccccccc} & & & & P_{x_j} & & \\ & & & & \downarrow & & \\ 0 & \longrightarrow & J & \longrightarrow & H^*(B_G,k) & \longrightarrow & H_G^*(X,k) \\ & & & & \downarrow & & \downarrow \\ & & 0 & \longrightarrow & H^*(B_{H_j},k) & \longrightarrow & H_{\tilde{H}_j}^*(X,k) \\ & & & & \downarrow & & \\ & & & & 0 & & \end{array}$$

Hence,

$$J \subseteq P_{x_j} \subseteq P_j$$

and the minimality of P_j clearly implies that $P_{x_j} = P_j$ which proves (i).

(ii) Let $Y^j = F(H_j, X)$, $j = 1, 2, \ldots, a$. Since H_j is maximal in $\mathcal{O}^0(X)$, it is easy to see that the induced action of $\tilde{H}_j = G/H_j$ on Y^j is almost free. Therefore

$$Y_G^j = E_G \times_G Y^j = (E_{H_j} \times E_{\tilde{H}_j}) \times_{(H_j \times \tilde{H}_j)} Y^j = B_{H_j} \times Y_{\tilde{H}_j}^j$$

and

$$H^*(Y_{\tilde{H}_j}^j, k) \cong H^*(Y^j/G; k).$$

Hence

$$H_G^*(X,k)_{P_j} \cong H_G^*(Y^j,k)_{P_j} \cong H^*(Y^j/G,k) \otimes_k H^*(B_{H_j},k)_{P_j}. \quad \square$$

Added in proof: In the general case that X is not necessary compact, one simply let J = Annahilator of $H_G^*(X,k)$, then Theorem (IV.6) holds in general.

Example (1). As a simple application of the above theorem, let us consider the case $\underline{G = T^r, X \sim_{\mathbb{Q}} S^n \text{ and } F = \emptyset}$:

Since the E_2-term of Serre spectral sequence of $X_G \to B_G$ consists of only two lines, $E_2^{*,0} = H^*(B_G,k) \otimes 1$ and $E_2^{*,n} = H^*(B_G,k) \otimes \{x\}$, it is clear that the generator $x \in H^n(X,k)$ is transgressive, and $E_{n+2} = E_\infty$. Hence the kernel, J, of the edge homorphism $H^*(B_G,k) \to H^*(X_G,k)$ is exactly the ideal generated by the transgression of x, $d_{n+1}(x) \in H^{n+1}(B_G,k)$. Then, the above theorem simply implies that

$$\alpha = d_{n+1}(x) = w_1^{k_1} w_2^{k_2} \cdots w_a^{k_a}$$

factorizes into the product of "linear polynomials" in $H^*(B_G,k) \cong k[t_1, \ldots, t_r]$ and $\{H_j = w_j^\perp, j = 1, 2, \ldots, a\}$ are exactly those maximal elements of $\mathcal{O}^0(X)$ and moreover

$$F(H_j, X)/G \sim_{\mathbb{Q}} CP^{(k_j - 1)}, \quad F(H_j, X) \sim_{\mathbb{Q}} S^{(2k_j - 1)}.$$

(2) Next, let us consider the case $\underline{X \sim_k S^n \text{ but } F = S^r \neq \emptyset}$:

In this case, $H_G^*(X,F)$ is a torsion module. Again, the E_2-term of Serre spectral sequence, $E_2 = H^*(B_G : k) \otimes H^*(X, F; k)$, consists of two lines. Therefore, it is not difficult to see that

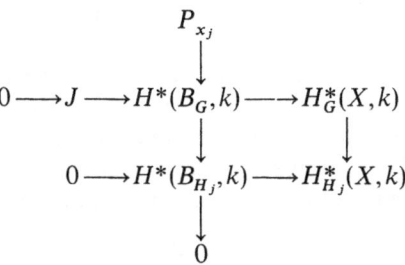

$$d_{(n-r)}(x) = \alpha \cdot (\partial f) \neq 0, \qquad \alpha \in H^{(n-r)}(B_G, k),$$

is the only non-zero differential, where x, ∂f are the generators of $H^n(X,F;k)$ and $H^{r+1}(X,F;k)$ respectively. Hence, we have that $H_G^*(X,F;k)$ is generated by $\{\partial f\}$ as a $H^*(B_G,k)$-module and the annahilator of ∂f is exactly the principal ideal $(\alpha) \subseteq H^*(B_G,k)$. Then, almost the same proof will show that the generator α again factorizes into the product of linear polynomials, namely,

$$\alpha = w_1^{k_1} \cdot w_2^{k_2} \cdots w_a^{k_a}$$

and $\{H_j = w_j^{\perp}; j=1,2,\ldots,a\}$ are correspondingly the maximal elements of $\mathcal{O}^0(X-F)$ with $F(H_j,X) \sim_k S^{(2k_j+r)}$.

Summarizing the above two cases, we get, in particular, the important Borel-Formula for topological actions of elementary abelian groups on homology spheres:

Theorem (IV.7) (A. Borel). *Let G be a torus (resp. p-torus) and $X \sim_k S^n$. Let $n(H)$ be the integer such that $F(H,X) \sim_k S^{n(H)}$, and $r = n(G)$. Then*

$$(n-r) = \sum_H (n(H) - r)$$

where H runs through the subtori (resp. sub-p-tori) of corank 1.

Theorem (IV.7)′. *Let G be a torus (resp. p-torus) and X be a k-cohomology manifold of dimension n. Let $x \in X$ be a fixed point and assume that x has a countable fundamental system of neighborhoods. Let $n(H)$ be the cohomology dimension of $F(H,X)$ at x and $r = n(G)$. Then, again*

$$(n-r) = \sum_H (n(H) - r)$$

where H runs through subtori (resp. sub-p-tori) of corank 1.

Remarks. (i) Theorem (IV.7) is simply the following statement about degrees:

$$\begin{aligned}\deg(\alpha) = (n-r) &= \deg(w_1^{k_1} \cdot w_2^{k_2} \cdots w_a^{k_a}) \\ &= \sum 2k_j = \sum_{H_j} (n(H_j) - r).\end{aligned}$$

(ii) Geometrically, one may view Theorem (IV.7)′ as a kind of "*transversality*" among the F-varieties when the G-space X is a cohomology manifold. We shall further explore the geometric significance as well as applications of Theorem (IV.7)′ in the next chapter.

(C) *Ideal Theoretical Invariants of Equivariant Cohomology*

With the results of the above discussion as our prototype, let us proceed to a general investigation of the geometric meaning of annahilating ideals in the structure of equivariant cohomology. To begin with, let us introduce the following notations:

$$\text{(i)} \begin{cases} R_G = H^*(B_G,k) & \text{if} \quad G = T^r \quad \text{or} \quad \mathbb{Z}_2^r \\ R_G = k[t_1,\ldots,t_r] \subseteq H^*(B_G,k) & \text{if} \quad G = \mathbb{Z}_p^r, p \neq 2. \end{cases}$$

(ii) For a closed subgroup $K \subseteq G$, let P_K be the kernel of $R_G \to R_K$, which is clearly the prime ideal with the Lie algebra of K as its variety.

$$\text{(iii)} \begin{cases} \text{For a given closed subgroup } K \subseteq G,\ X^K = F(K,G) = \{x \in X;\ G_x \supseteq K\}, \\ \text{For a given prime ideal } P \subseteq R_G,\ X^P = \{x \in X;\ P_{G_x} \subseteq P\}. \end{cases}$$

(iv) The annahilator of an R_G-submodule $M \subseteq H_G^*(X, Y; k)$ is defined by $\operatorname{ann} M = \{a \in R_G;\ a \cdot M = 0\}$. In case M is generated by $\{m_1, \ldots, m_l\}$, then we may also write $\operatorname{ann} M$ as $\operatorname{ann}(m_1, \ldots, m_l)$.

(v) For a given closed subgroup $K \subseteq G$ and an R_G-module $M \subseteq H_G^*(X, Y; k)$ we shall denote the image of M in $H_G^*(X^K, Y^K; k)$ by ${}^K M$.

(vi) Let Y be a closed invariant subspace of the G-space X, and η be a given element of $H_G^*(Y)$. Let

$$I_\eta(X, Y) = \{a \in R_G;\ a \cdot \eta \in \operatorname{Im}(H_G^*(X) \to H_G^*(Y))\}.$$

It follows from the H_G^*-exact sequence of the pair (X, Y) that

$$I_\eta(X, Y) = \operatorname{ann}(\partial \cdot \eta),\ \partial\eta \in H_G^*(X, Y).$$

$$\text{(vii)} \qquad I_\eta^K(X, Y) = I_\eta(X^K \cup Y, Y).$$

In Theorem (IV.6), we studied the ideal $J = \operatorname{ann}(1) = \operatorname{ann} H_G^*(X)$ and succeeded in determining the geometric meaning of the prime components of J. Hence, it is reasonable to look into the Noether-Lasker decomposition of the above ideals, $\operatorname{ann} M$ or I_η, and try to determine the geometric meaning of their primary components. First, let us point out some simple basic facts which play a crucial rôle in the proof of Theorem (IV.6) as well as its various generalizations that we shall discuss in this section.

Proposition 3. *In the case G is a torus (resp. p-torus), the correspondence assigning each subtorus K (resp. sub-p-torus) to the ideal $P_K = \operatorname{Ker}(B_G \to B_K)$ gives a bijection between the set of subtori in G and the set of linear ideals in R_G. Moreover, to a given prime ideal $P \subseteq R_G$, there exists a unique minimal subtorus $K \subseteq G$ satisfying $P_K \subseteq P$.*

Proof. The first assertion is a direct consequence of the fact that the variety of P_K, $V(P_K)$, is exactly the Lie algebra of K and there is obvious bijection between linear subspaces of $\mathfrak{g}$, i.e., Lie subalgebras of $\mathfrak{g}$, and subtori of G. The second assertion follows from the fact that there exists a unique linear subspace spanned by the variety of P, $V(P)$, and K is just the subtorus with that linear subspace as its Lie algebra.

Proposition 4. *For a given prime ideal $P \subseteq R_G$, the homomorphism*

$$H_G^*(X^P, Y^P) \to H_G^*(X^P, Y^P)_P$$

is always injective.

Proof. Let K be the minimal subtorus with $P_K \subseteq P$. Then it is clear that $X^P = X^K$, $Y^P = Y^K$. Hence, we need only to show that the following composition is injective:

$$H_G^*(X^K, Y^K) \to H_G^*(X^K, Y^K)_P \to H_G^*(X^K, Y^K)_{P_K}.$$

Since G is a torus and K is a subtours, there exists K' such that $G = K \times K'$ and $E_G = E_K \times E_{K'}$. Hence

$$X_G^K = (E_K \times E_{K'}) \times_G X^K = B_K \times (E_{K'} \times_{K'} X^K) = B_K \times X_{K'}^K$$

and $Y_G^K = B_K \times Y_{K'}^K$ with respect to the induced K' action on X^K, Y^K. Therefore, by Kunneth formula, we have the commutative diagram:

$$\begin{array}{ccc} H_G^*(X^K, Y^K) \cong & H^*(B_K) \otimes H_{K'}^*(X^K, Y^K) \\ \downarrow & \downarrow \\ H_G^*(X^K, Y^K)_{P_K} \cong & H^*(B_K)_{P_K} \otimes H_{K'}^*(X^K, Y^K) \end{array}$$

where the right vertical map is the tensoring of $H^*(B_K) \to H^*(B_K)_{P_K}$ with the identity map which is obviously injective. $\square$

Combine the above propositions with the localization theorem, we get

Proposition 5. *Let $P \subseteq R_G$ be a prime ideal and K be the minimal subtours with $P_K \subseteq P$. Then*

$$(\operatorname{ann} M)_P \cap R_G = \operatorname{ann} {}^K M.$$

Proof. It follows from the following commutative diagram that M_P is generated by ${}^K M$ as an $(R_G)_P$-module;

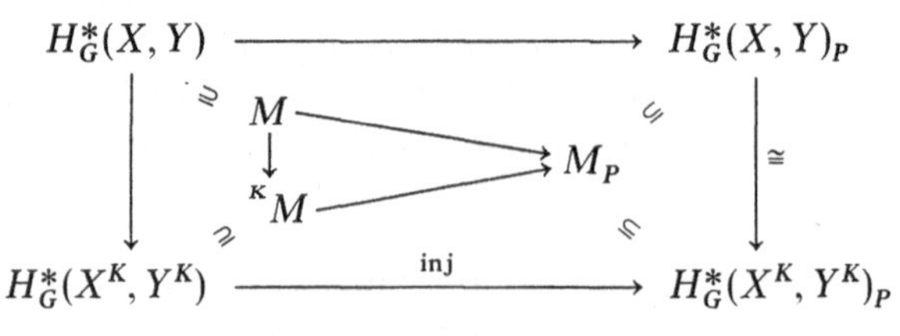

Hence

$$(\operatorname{ann} M)_P \cap R_G = \operatorname{ann}(M_P) \cap R_G = \operatorname{ann} {}^K M.$$

Next, let us recall some basic facts of Noether-Lasker primary decomposition in terms of localization as follows:

Let $R = k[t_1, \ldots, t_r]$ be a polynomial ring over a field k. Then every ideal $I \subseteq R$ has a reduced primary decomposition:

$$I = q_1 \cap \cdots \cap q_m; \quad (q_i \text{ primary ideals}).$$

Let P be an arbitrary prime ideal and $P_i = \sqrt{q_i}$. Then

$$I_P \cap R = \bigcap \{q_i; P_i \subseteq P\},$$

Therefore, those prime ideals $\{P_1, \ldots, P_m\}$ belonging to I can be characterized by

$$P = P_i \quad \text{if and only if} \quad I_P \cap R \subsetneq \bigcap \{I_{P'} \cap R); P' \subsetneq P\}$$

and the primary component q_i with *minimal* P_i is given by

$$q_i = I_{P_i} \cap R.$$

As a straightforward generalization of Theorem (IV.6), let us now prove the following:

Theorem (IV.8) [S4]. *Let M be a finitely generated R_G-submodule of $H_G^*(X, Y)$ and*

$$\operatorname{ann} M = q_1 \cap \cdots \cap q_m$$

be a reduced primary decomposition of $\operatorname{ann} M$. *Then*

(i) *the prime ideals* $\{\sqrt{q_1}, \ldots, \sqrt{q_m}\}$ *belonging to* $\operatorname{ann} M$ *are all linear and* $\sqrt{q_i} = P_{K_i}$ *where* K_i *are those subtori characterized by the following properties*

$$\operatorname{ann}^K M \subsetneq \bigcap \{\operatorname{ann}^{K'} M; K' \supsetneq K\}, \quad \textit{and moreover,}$$

(ii) P_{K_i} *is minimal in* $\{\sqrt{q_1}, \ldots, \sqrt{q_m}\}$ *if and only if* K_i *is maximal among all subtori with* ${}^K M \neq 0$, *and in this case,*

$$q_i = \operatorname{ann}^{K_i} M.$$

Proof. (i) Let $\operatorname{ann} M = q_i \cap \cdots \cap q_m$ be a reduced primary decomposition of $\operatorname{ann} M$ and $\{P_i = \sqrt{q_i}; i = 1, \ldots, m\}$ be the prime ideals belonging to $\operatorname{ann} M$. Then P_i are characterized by

$$P = P_i \Leftrightarrow (\operatorname{ann} M)_P \cap R_G \subsetneq \bigcap \{(\operatorname{ann} M)_{P'} \cap R_G;\ P' \subsetneq P\}.$$

However, it follows immediately from proposition 5 that

$$(\operatorname{ann} M)_P \cap R_G = (\operatorname{ann} M)_{P_K} \cap R_G = \operatorname{ann}^K M$$

where K is the minimal subtorus with $P_K \subseteq P$. Hence, we may replace all the prime ideals in the above characterization by linear ideals and (i) follows immediately.

(ii) Now, suppose $P_1, \ldots, P_l$ are those minimal primes in $\{P_1, \ldots, P_m)$. Then $P_1, \ldots, P_l$ are exactly those minimal prime ideals *containing* $\operatorname{ann} M$, or equivalently $(\operatorname{ann} M)_P \cap R_G \neq R_G$. Hence, again by Proposition 5, $\{P_i = P_{K_i}, i = 1, \ldots, l\}$ are exactly those prime ideals which are minimal with respect to the condition

$(\operatorname{ann} M)_{P_K} \cap R_G = \operatorname{ann}^K M \neq R_G$, or equivalently ${}^K M \neq 0$. That is, $K_1, \ldots, K_l$ are those maximal subtori with ${}^K M \neq 0$. □

Examples and preliminary applications. (1) Suppose $Y \subseteq X$ is an invariant subspace and $\eta \in H_G^*(X)$. Then we have

$$I_\eta(X, Y) = \{a \in R_G;\, a\eta \in \operatorname{Im}(H_G^*(X) \to H_G^*\, Y)\} = \operatorname{ann}(\partial\eta)$$

and we may apply the above theorem to $I_\eta(X, Y)$ to get the following corollary:

Corollary. *Let $I_\eta(X, Y) = q_1 \cap \cdots \cap q_m$ be a reduced primary decomposition. Then $P_i = \sqrt{q_i} = P_{K_i}$ are linear prime ideals and K_i are characterized by the following properties*

$$I_\eta^K(X, Y) \subsetneq \bigcap \{I_\eta^{K'}(X, Y);\, K' \supsetneq K\}, \quad \textit{and moreover,}$$

P_{K_i} is minimal in $\{P_1, \ldots, P_m\}$ if and only if K_i is maximal among all subtori with $I_\eta^K(X, Y) \neq R_G$, and in this case,

$$q_i = I_\eta^{K_i}(X, Y).$$

Proof. Recall that $I_\eta^K(X, Y) = I_\eta(X^K \cup Y, Y)$, and it follows easily from the commutative diagram

$$\begin{array}{ccccccc}
H_G^*(X) & \longrightarrow & H_G^*(Y) & \xrightarrow{\partial} & H_G^*(X, Y) & \searrow & \\
\downarrow & & \downarrow & & \downarrow & & \\
H_G^*(X^K \cup Y) & \longrightarrow & H_G^*(Y) & \xrightarrow{\partial} & H_G^*(X^K \cup Y, Y) & \xrightarrow[\cong]{} & H_G^*(X^K, Y^K)
\end{array}$$

that $I_\eta(X, Y) = \operatorname{ann}(\partial\eta)$ and $I_\eta^K(X, Y) = \operatorname{ann}^K(\partial\eta)$. Hence, the above corollary follows from theorem (IV.8) by direct specification. □

(2) In application, the following fact is often useful.

Proposition 6. *Suppose that the quotient module of $H_G^*(X)$ by it R_G-torsion submodule is a free R_G-module. Then $I_\eta(X, F)$ is principal for every $\eta \in H_G^*(F)$, (where $F = F(G, X)$) and its generator splits into product of linear polynomials.*

Proof. Let N be the R_G-torsion submodule of $H_G^*(X)$ and M be a lifting of the free module H_G^*/N back into $H_G^*(X)$. Then

$$H_G^*(X) = M \oplus N \quad (\text{as } R_G\text{-modules})$$

and $M \subseteq H_G^*(X) \to H_G^*(F)$ maps M injectively into $H_G^*(F)$. Since

$$S^{-1}M \cong S^{-1}H_G^*(F), \qquad S = R_G - \{0\};$$

it is clear that any element $\eta \in H_G^*(F)$ can be expressed in terms of a basis $\{m_i\}$ of the free module M with *fractional* coefficients, namely

$$\eta = \sum \left(\frac{b_i}{a_i}\right) m_i = \frac{1}{a} \sum c_i m_i$$

where $a_i, b_i, c_i, a \in R_G$ and $\{c_i\}$ *have no common divisor.* Then, it follows easily from the unique factorization property of R_G that $I_\eta(X,F)=(a)$ is the principal ideal generated by a.

The fact that a splits into product of linear polynomials is a direct consequence of the above corollary of Theorem (IV.8).

Corollary. *Let* $X_1=\{x \in X;\ \text{corank } G_x \leqslant 1\}$. *If* $H_G^*(X)$ *is a free* R_G*-module, then the following map is injective:*

$$H_G^*(X,F) \longrightarrow H_G^*(X_1,F).$$

Proof. Since $H_G^*(X)$ is free and $H_G^*(X,F)$ is torsion, the long exact sequence of the pair (X,F) becomes a short exact sequence, namely,

$$0 \longrightarrow H_G^*(X) \longrightarrow H_G^*(F) \xrightarrow{\delta} H_G^*(X,F) \longrightarrow 0.$$

To an arbitrary element $\xi \in H_G^*(X,F)$, there exist $\eta \in H_G^*(F)$ such that $\xi=\partial\eta$. Hence it follows from Proposition 6 that $\operatorname{ann}(\xi)=I_\eta(X,F)$ is principal and there exists subtorus K of corank 1 such that ξ does not map to zero in $H_G^*(X^K,F)$. Because $X^K \subseteq X_1$, ξ does not map to zero in $H_G^*(X_1,F)$ either. But ξ is arbitrary, the map $H_G^*(X,F) \to H_G^*(X_1,F)$ must be injective. □

Remark. Later on, we shall apply the above result to deduce a formula of Golber [G 4] which is a second order formula of the Borel types [cf. § 1, Ch. VIII].

(3) Suppose X is a G-space with disconnected fixed point set F; let F^1, F^2 be two connected components of F and $x_1 \in F_1$, $x_2 \in F_2$. Let $Y=\{x_1,x_2\}$, and $\eta \in H^0(Y) \subseteq H_G^*(Y)$ be the element which has value 1 on x_1 but 0 on x_2. Then, it is not difficult to see that the *minimal prime ideals* belonging to $I_\eta(X,Y)$ correspond exactly to the *generic isotropy subgroups* of those *minimal* F^0-varieties (resp. F-varieties) connecting F^1 and F^2.

Similarly, let $Y=F=F^1+F^2+\cdots+F^s$ and

$$\eta_1=(1,0,\ldots,0) \in H^0(F)=\textstyle\sum_{i=1}^s H^0(F^i) \subseteq \sum_{i=1}^s H^*(F^i) \subseteq \sum_{i=1}^s H_G^*(F^i).$$

Then, it is not difficult to see that the *minimal prime ideals* belonging to $I_{\eta_1}(X,F)$ correspond to the *generic isotropy subgroups* of those minimal F^0-varieties (resp. F-varieties) *connecting* F^1 with other components. Suppose, in particular, the quotient of $H_G^*(X)$ by its R_G-torsion submodule is a free R_G-module. Then it follows from Proposition 6 that $I_{\eta_1}(X,F)$ is principal and hence the generic isotropy subgroups of those minimal F^0-varieties (resp. F-varieties) connecting F^1 to other components must be of corank 1 (i.e., the generator of $I_{\eta_1}(X,F)$ factorizes into the product of linear polynomials).

Example. As a simple, concrete application of the above ideal theoretical invariants and Theorem (IV.8), let us consider the case

$$G=T^r(r\geqslant 2) \quad \text{and} \quad X\sim_k QP^n, \quad \text{i.e.,} \quad H^*(X,k)\cong k[\xi_0]/\xi_0^{n+1}, \quad \deg\xi_0=4.$$

Again, the Serre spectral sequence of the fibration $X\to X_G\to B_G$ has no non-zero differentials because $E_2=H^*(B_G)\otimes_k H^*(X)$ has no odd degree elements. Hence $E_2=E_\infty$ and $H_G^*(X)\cong H^*(B_G)\otimes_k H^*(X)$ as $H^*(B_G)$-module. Let ξ be a lifting of $\xi_0\in H^4(X)$ into $H_G^4(X)$, then $\{1,\xi,\xi^2,\dots,\xi^n\}$ forms an R_G-module basis of $H_G^*(X)$ and

$$H_G^*(X)\cong R_G[\xi]/\langle f(\xi)\rangle \quad (\text{as an } R_G\text{-algebra}),$$

where $f(\xi)=(\xi^{n+1}+c_1\xi^n+c_2\xi^{(n-1)}+\dots+c_{n+1})$, $c_i\in H^{4i}(B_G)$. It follows from Theorem (IV.1) that $f(\xi)$ splits into a product of "linear" factors in $R_G[\xi]$, namely,

$$f(\xi)=(\xi-\alpha_1)^{k_1}(\xi-\alpha_2)^{k_2}\cdots(\xi-\alpha_s)^{k_s}.$$

And correspondingly, the fixed point set F consists of exactly s connected components $F^1,F^2,\dots,F^s$; such that

$$H^*(F^j,k)\cong k[\eta_j]/\langle\eta_j^{k_j}\rangle, \quad \deg\eta_j=2 \quad \text{or} \quad 4.$$

Let $\varphi_j\colon H_G^*(X)\to H_G^*(F^j)$ and $\varphi\colon H_G^*(X)\to H_G^*(F)=\sum_{j=1}^s H_G^*(F^j)$. Then it is clear that

$$\begin{aligned}
&\varphi_j(\xi)=\alpha_j \quad \text{if} \quad F^j\sim_k pt, \quad \text{i.e.,} \quad k_j=1,\\
&\varphi_j(\xi)=\eta_j+\alpha_j \quad \text{if} \quad F^j\sim_k QP^{(k_j-1)}, \quad \text{i.e.,} \quad k_j>1 \quad \text{and} \quad \deg\eta_j=4,\\
&\varphi_j(\xi)=\eta_j^2+\beta_j\eta_j+\alpha_j, \quad \text{if} \quad F^j\sim_k CP^{(k_j-1)}, \quad \text{i.e.,} \quad k_j>1 \quad \text{and} \quad \deg\eta_j=2
\end{aligned}$$

and

$$\varphi(\xi)=(\varphi_1(\xi),\varphi_2(\xi),\dots,\varphi_s(\xi))\in H_G^4(F)=\sum_{j=1}^s H_G^4(F^j).$$

Let $f^j=(0,\dots,0,\eta_j^{(k_j-1)},0,\dots,0)\in H^*(F)\subseteq H_G^*(F)$ and we shall compute the ideal $I_{f^j}(X,F)$ explicitly as follows:

Since $H_G^*(X)$ is a free R_G-module, $I_j=I_{f^j}(X,F)$ is a principal ideal, say generated by a_j. Suppose $g(\xi)\in H_G^*(X)$ is an element with $\varphi_i(g(\xi))=0$ for all $i\neq j$ and $\varphi_j(g(\xi))=a\cdot f^j$. It follows from the condition $\varphi_i(g(\xi))=0$ that $g(\xi)$ must be divisible by $(\xi-\alpha_i)^{k_i}$, and the condition $\varphi_j(g(\xi))=a\cdot f^j=a\cdot\eta_j^{(k_j-1)}$ implies that $\varphi_j(g(\xi)(\xi-\alpha_j))=0$, i.e., $g(\xi)$ is divisible by $(\xi-\alpha_j)^{(k_j-1)}$. Hence

$$g(\xi)=\left(\prod_{i\neq j}(\xi-\alpha_i)^{k_i}\right)\cdot(\xi-\alpha_j)^{(k_j-1)}\cdot\tilde{g}(\xi).$$

Let $g_0(\xi)=\left(\prod_{i\neq j}(\xi-\alpha_i)^{k_i}\right)\cdot(\xi-\alpha_j)^{(k_j-1)}$. Then it is easy to see that

$$\varphi(g_0(\xi))=(0,\dots,0,a_jf^j,0,\dots,0)$$

where $a_j = \prod_{i \neq j}(\alpha_j - \alpha_i)^{k_i}$ if $\deg \eta_j = 4$ and $a_j = \prod_{i \neq j}(\alpha_j - \alpha_i)^{k_i} \cdot \beta_j^{(k_j - 1)}$ if $\deg \eta_j = 2$, and moreover $a = a_j \cdot$ [constant term of $\varphi_j(\tilde{g}(\xi))$]. As a direct consequence of the above computation and Theorem (IV.8), we have that $(\alpha_j - \alpha_i)$ *splits into product of two "linear" polynomials for all* $i \neq j$. A complete determination of the cohomological invariants of torus actions of spaces of QP^n-type was given by Hsiang and Su [H21]. We shall discuss their result in Chapter VI.

§ 3. A Splitting Theorem for Poincaré Duality Spaces

Recall that a Poincaré duality space, X, of formal dimension n over k satisfies the following conditions:

(i) $H^n(X,k) = k$ and $H^i(X,k) = 0$ for $i > n$,

(ii) $H^j(X,k) \times H^{(n-j)}(X,k) \xrightarrow{\text{"}\cup\text{"}} k$ is perfect for all $0 \leqslant j \leqslant n$.

It was conjectured by Su that each connected component of the fixed point set of torus group (resp. p-torus) on a P.D. space over $\mathbb{Q}$ (resp. $\mathbb{Z}_p$) is again a P.D. space over $\mathbb{Q}$ (resp. $\mathbb{Z}_p$). Bredon [B16] proved the above conjecture under the very restrictive assumption that X is totally non-homologous to zero in the fibration $X \to X_G \to B_G$. Recently, T. Chang and T. Skjelbred proved the above conjecture in its full generality in their theses [C2]. Furthermore, they also proved in their theses the following conjecture of Hsiang: Let G be a torus (resp. a p-torus) and X be a P.D. G-space over $\mathbb{Q}$ (resp. $\mathbb{Z}_p$) with non-empty fixed point set F. For a connected component F^j of F, the *local geometric weight system at* F^j is defined in terms of F^0-variety (resp. F-variety) structure around F^j as follows:

Let Y_i be those F^0-varieties containing F^j and with *corank one generic isotropy subgroups*, say, denoted by $w_i^{\perp}$. Let $\dim Y_i$, $\dim F^j$ be the formal dimension of the P.D. spaces Y_i, F^j respectively. Then the local geometric weight system at F^j is defined to be

$$\Omega_{F^j} = \{w_i, m_i\}$$

where the multiplicity of w_i is given by $m_i = (\dim Y_i - \dim F^j)$.

Conjecture. Let x, f^j be the fundamental class of X and the connected component, F^j, of the fixed point set F respectively. Then the ideal $I_{f^j}(X,F)$ is the principal ideal generated by the product of local weights, namely,

$$a^j = w_1^{d_1} \cdots w_l^{d_l}$$

where the local weight system at F^j is: $\Omega_{F^j} = \{w_i, m_i; i = 1, \ldots, l\}$ with $m_i = 2d_i$ if $p \neq 2$, $m_i = d_i$ if $p = 2$. Moreover, there exists a suitable lifting, x_j, of x into $H_G^*(X)$ such that the restriction of x_j to $H_G^*(F^j)$ is exactly $a_j \cdot f^j$ but the restrictions of x_j to $H_G^*(F^i)$ are zero for all $i \neq j$.

In this section, we shall give their proof of the above two conjectures:

Theorem (IV.9) [C2]. *Let G be a torus (resp. p-torus) and X be a P.D. G-space over* $\mathbb{Q}$ *(resp.* $\mathbb{Z}_p$*). Then the connected components of the fixed point set F are also P.D. spaces over* $\mathbb{Q}$ *(resp.* $\mathbb{Z}_p$*).*

Theorem (IV.10) [C2]. *Let G be a torus (resp. p-torus) and X be a P.D. G-space over* $\mathbb{Q}$ *(resp.* $\mathbb{Z}_p$*). Let* $\Omega_{F^j}=\{w_i,m_i\}$ *be the local weight system at the connected component,* F^j*, of the fixed point set F, and* f^j *be the fundamental class of* F^j*. Then* $I_{f^j}(X,F)$ *is the principal ideal generated by*

$$a_j=\prod w_i^{d_i} \quad (d_i=m_i \text{ if } p=2 \text{ and } d_i=\frac{m_i}{2} \text{ otherwise}).$$

Proof. We shall only give here a complete proof for the case $p=0,2$; and refer the reader to [C2, § 4] for the necessary minor modifications due to the fact that $R_G \subsetneq H^*(B_G,k)$ for odd primes. We shall need the following two simple lemmas.

Lemma 1. *Let* $X\to M\to B$ *be fibration where B is a closed, orientable piecewise linear manifold of dimension b and X is a P.D. space (over k) of dimension n. Suppose the local system* $\{H^n(X,k)\}$ *is constant over B. Then M is also a P.D. space (over k) of dimension* $(n+b)$.

Proof. Using open cells in B and the Mayer-Vietories sequence, the usual proof of Poincaré duality of B, with simple modification, will also give the proof of Poincaré duality of M. □

Lemma 2. *Assume that* $H^n(X,k)=k$ *and* $H^q(X,k)=0$ *for* $q>n$*. Then X is a P.D. space of dimension n if and only if, for every* $c\in H^q(X,k)$, $q<n$, $c\neq 0$*, there is* $c_1\in H^p(X,k)$ *with* $p>0$ *and* $c\cdot c_1\neq 0$.

We shall now embark on the proof of Theorem (IV.9) and (IV.10). The main idea of the proof is roughly as follows: One may replace the base space B_G by a sufficiently high dimensional approximation, say $(CP^N)^r\subseteq(CP^\infty)^r=B_{T^r}$ and $(RP^N)^r\subseteq(RP^\infty)^r=B_{\mathbb{Z}_2^r}$, which is a manifold. Let F^j be a connected component of the fixed point F and q^j be a point of F^j. Let $M(X)$, $M(F^j)$, $M(q^j)$ be the restrictions of X_G, F_G^j and $\{q^j\}_G$ over such an approximation $B\subseteq B_G$ respectively. Then, by Lemma 1, $M(X)$ is a P.D. space of formal dimension $\dim M(X) = (\dim X+\dim B)$. Let $\xi_j\in H^n(M(X))$ be the dual of $[M(q^j)]$. We shall show that ξ_j is a lifting of the fundamental cohomology class $\xi\in H^n(X)$ into $H^n(M(X))$ such that

$$\varphi_j^*(\xi_j)=a_j\cdot f^j \quad\text{and}\quad \varphi_i^*(\xi_j)=0 \quad\text{for}\quad i\neq j$$

where φ_i^* are the restriction homomorphisms induced by $M(F^i)\subseteq M(X)$, and moreover, f^j is the fundamental cohomology class of F^j and a_j is the generator of $I_{f^j}(X,F)$.

(i) When M is a P.D. space over k, we shall denote the fundamental homology class by $[M]$ which is a chosen generator of $\mathrm{Hom}(H^n(M,k),k)$. By P.D. of

$M(X)$, there exists a unique $\xi_j \in H^n(M(X))$ such that

$$\langle x\xi_j, [M(X)]\rangle = \langle x, [M(q^j)]\rangle \quad \text{for all} \quad x \in H^*(M(X)).$$

Since $M(q^j) \subseteq M(X)$ is a cross-section of the fibration $X \to M(X) \to B$, it is not difficult to see that $\xi_j \in H^n(M(X))$ is a lifting of the fundamental class $\xi \in H^n(X)$. Thus ξ_j generates a free $H^*(B)$-submodule in $H^*(M(X))$.

(ii) Recall that $H_G^*(X,F)$ is a finitely generated torsion R_G-module. Hence there is $a \in (R_G - \{0\})$ such that $a \cdot H_G^*(X,F) = 0$. For a given element $0 \neq c \in H^q(F^j)$, we define a homomorphism $\gamma: H^*(M(F^j)) \to k$ such that $\gamma(H^i(B) \times H^l(F^j)) = 0$ if $(i,l) \neq (\dim B, q)$ and $\gamma(b \times c) = 1$ for a chosen fundamental cohomology class, b, of B. Then, by P.D. of $M(X)$, there exists a unique $D(c) \in H^{(n-q)}(M(X))$ such that

$$\langle x \cdot D(c), [M(X)]\rangle = \gamma(x \,|\, M(F^j)), \quad \text{for all} \quad x \in H^*(M(X)).$$

Consider c as an element of $H_G^*(F)$ by $c \in H^*(F^j) \subseteq H^*(F) \subseteq H_G^*(F)$. Then, it follows from the fact $a \cdot H_G^*(X,F) = 0$ that there exists $\tilde{c} \in H_G^*(X)$ with $\varphi^*(\tilde{c}) = a \cdot c$ (where φ^* is the restriction homomorphism induced by $M(F) \subseteq M(X)$). Therefore, we have

$$\begin{aligned}\langle x\tilde{c} \cdot D(c), [M(X)]\rangle &= \gamma(x\tilde{c} \,|\, M(F^j)) = \gamma(xac \,|\, M(F^j)) = \langle xa, [M(q^j)]\rangle \\ &= \langle xa\xi_j, [M(X)]\rangle \text{ for all } \quad x \in H^*(M(X)),\end{aligned}$$

and consequently, by P.D. of $M(X)$, $\tilde{c} \cdot D(c) = a \cdot \xi_j$.

(iii) Take $c_0 = 1 \in H^*(F^j)$. We have $\varphi(\tilde{c}_0) = (0, \ldots, a, \ldots, 0)$, i.e., $\varphi_i^*(\tilde{c}_0) = 0$ for $i \neq j$ and $\varphi_j^*(\tilde{c}_0) = a$. Hence, it follows from $\tilde{c}_0 D(c_0) = a\xi_j$ that

$$a\varphi_i^*(\xi_j) = \varphi_i^*(\tilde{c}_0 \cdot D(c_0)) = 0 \quad \text{for} \quad i \neq j,$$

and $\varphi_j^*(\xi_j) \neq 0$ because ξ_j is R_G-torsion-free. Now, let $H^l(F^j)$ be the top dimensional non-zero cohomology group and c_1 be a non-zero element of $H^l(F^j)$. Then $\tilde{c}_1 \cdot D(c_1) = a \cdot \xi_j$ implies that $\varphi^*(\tilde{c}_1 \cdot D(c_1)) = (0, \ldots, 0, \varphi_j^*(\tilde{c}_1 \cdot D(c_1)), 0, \ldots, 0)$ and $\varphi_j^*(\tilde{c}_1 \cdot D(c_1)) = a \cdot c_1 \, \varphi_j^*(D(c_1)) = a \cdot \varphi_j^*(\xi_j)$ therefore, due to the fact c_1 is top dimensional, we have

$$c_1 \cdot \varphi_j^*(D(c_1)) = c_1 \cdot a_j = \varphi_j^*(\xi_j)$$

where $a_j =$ the restriction of $D(c_1)$ to $M(q^j) \in H^*(B_G)$. Hence $H(F^j, k) = k$ and generated by c_1. Next, let c be an arbitrary non-zero element of $H^*(F^j)$. Then, again, $\tilde{c} \cdot D(c) = a\xi_j$ implies that

$$c \cdot \varphi_j^*(D(c)) = \varphi_j^*(\xi_j) = a_j \cdot c_1 \neq 0$$

which, by Lemma 2, shows that F^j is a P.D. space with $c_1 = f^j$ as a fundamental cohomology class.

(iv) Let $\psi: H_G^*(X) \to E_\infty^{*,n}$ be the projection onto the first subquotient of the filtered R_G-module $H_G^*(X)$, where $E_\infty^{*,n}$ is a free R_G-module generated by $\psi(\xi_j)$.

Let a' be an arbitrary element of $I_{f^j}(X,F)$. Then, by definition, there is $\xi' \in H_G^*(X)$ such that $\varphi^*(\xi') = a' \cdot f^j$. Hence, we have

$$\varphi^*(a' \cdot \xi_j - a_j \xi') = (a' \cdot a_j - a_j a') \cdot f^j = 0 \Rightarrow (a' \xi_j - a_j \xi')$$

is torsion.

But $E_\infty^{*,n}$ is a free R_G-module generated by $\psi(\xi_j)$, we have $\psi(a' \xi_j - a_j \xi') = 0$, so that

$$a' \psi(\xi_j) = a_j \psi(\xi') = a_j u(\xi_j) \Rightarrow a' = a_j u .$$

Hence $I_{f^j}(X,F)$ is principal and generated by a_j.

(v) It follows from Theorem (IV.8) that the generator a_j of $I_{f^j}(X,F)$ splits into a product of linear polynomials, say

$$a_j = \prod w_i^{d_i}$$

where $(w_i^{d_i}) = I_{f^j}(X^{H_i}, F)$ and $H_i = w_i^{\perp}$.

Let Y_i be the connected component of X^{H_i} containing F^j. Then it is obvious that

$$I_{f^j}(X^{H_i}, F) = I_{f^j}(Y_i Y_i \cap F) .$$

Now, by the above proof, Y_i is again a P.D. space. Hence, if we take the special case $X = Y_i$, the above proof clearly shows that $I_{f^j}(Y_i, Y_i \cap F) = (w_i^{d_i})$ and $\deg(w_i^{d_i}) = (\dim Y_i - \dim F^j)$. This completes the proof of Theorem (IV.9) and (IV.10).

Concluding remarks of Chapter IV. In concluding this chapter, let us add a couple of remarks which we hope will help to clarify the general perspective of the results of this chapter.

(i) Almost all the theorems of this chapter hold only for those elementary abelian compact groups (i.e., torus or p-torus), and fail miserably for all the other compact groups. In analyzing the proofs of theorems of this chapter, it is rather clear that the basic reason for such a radical and drastic difference between the geometric behaviors of actions of elementary abelian groups and those of other compact groups is the following crucial property that uniquely singles out elementary abelian groups from all the other compact topological groups. Namely, there is a *canonical bijection* between the set of connected subgroups (resp. subgroups) of a torus (resp. p-torus) G and the set of linear ideals of $H^*(B_G,k)$, $k = \mathbb{Q}$ (resp. $\mathbb{Z}_p$), given by assigning $K \subseteq G$ to the linear ideal:

$$P_K = \operatorname{Ker}\{H^*(B_G,k) \to H^*(B_K,k)\} .$$

[Notice that the variety of P_K is exactly the Lie algebra of K.]

Technically, the localization theorem then provides us with a convenient bridge to establish a tight relationship between isotropy subgroups and annihilating ideals.

(ii) As an important consequence of the Schur lemma, one has the fundamental fact that any complex representation of an elementary abelian group splits into the direct sum of one-dimensional complex representations. In the theory of characteristic classes of vector bundle, if the structural group can be reduced to an elementary abelian group, then the above *splitting at the geometric level* obviously gives us the splitting of characteristic classes. As pointed out in Chapter III, the equivariant cohomology theory can be viewed as a generalized version of characteristic class theory for bundles with general G-spaces as fibre. Hence, the geometric significance of those theorems of this chapter is the following: For general (i.e., not necessary linear) topological transformations of elementary abelian groups, one no longer has any kind of splitting at the geometric level. However, one can still prove several kinds of splitting at the characteristic class level, which seems to us to be the most crucial and fascinating fact in the whole theory of topological transformation groups. Actually, the linearity of various ideals as well as the splitting of various structural data that we proved in this chapter should be considered as the cohomological substitutes for the Schur lemma and we shall see later that they play a similar rôle in the cohomology theory of topological transformation groups as that of the Schur lemma in the theory of linear transformation groups.

(iii) In the discussion of this chapter, we have been consistently using the coefficient ring $\mathbb{Q}$ for the case $G=T^r$ and the coefficient ring $\mathbb{Z}_p$ for the case $G=\mathbb{Z}_p^r$. This setting is not just for technical convenience, but rather, this is the *only* setting that can possibly accommodate the type of theorems of this chapter. However, in the case $G=T^r$, one may sometimes obtain results with $\mathbb{Z}$-coefficient from those of $\mathbb{Q}$-coefficient by a straightforward application of Gauss lemma.

Chapter V. The Splitting Principle and the Geometric Weight System of Topological Transformation Groups on Acyclic Cohomology Manifolds or Cohomology Spheres

In this chapter, we apply the general theorems of Chapter IV to the important testing spaces of acyclic cohomology manifolds and cohomology spheres. Observe that, in the setting of topological transformation groups, there is a simple direct relationship between actions on acyclic cohomology manifolds and actions on cohomology spheres, which can be explained as follows. For a given action on a cohomology sphere X, there is a natural induced action on the cone of X, CX, which is an acyclic cohomology manifold. On the other hand, the restriction of an action on an acyclic manifold X to the complement of a fixed point x (if it exists) gives an action on the cohomology sphere $(X-\{x\})$. Hence, one need only treat one case and the corresponding result for the other case will follow automatically. In this chapter, we prefer to state the results for the case of acyclic cohomology manifolds because it is the directly applicable to the study of the local theory.

Traditionally, acyclic cohomology manifolds and cohomology spheres (e.g., $\mathbb{R}^n$ and S^n) are always regarded as the most important *testing spaces* for the study of transformation groups. The following are some of the special features that make them outstanding from the viewpoint of transformation groups. (i) The euclidean spaces as well as spheres are topologically simple on the one hand, and on the other hand, the abundant examples of linear actions and the equivariant embedding theorem of Mostow-Palais [M 8, P 1] fully demonstrate that they are rich both in variety and in complexity from the viewpoint of transformation groups. (ii) Technically, the existence of abundance *linear* examples provide us with invaluable guidance in correctly formulating problems as well as in guessing the right answers, and the simplicity of their topological structures makes all kinds of topological machinery such as cohomology theory and bundle theory easily applicable. (iii) Theoretically, the existence of *slices* [cf. § 3, Ch. I] reduces the local theory of compact transformation groups to the study of transformation groups on acyclic cohomology manifolds with non-empty fixed point set.

Recall that in the study of geometric behavior of transformation groups in the framework of modern topology, there are the following two natural settings: (i) topological actions of Lie groups (or more generally, topological groups) on topological manifolds (resp. topological spaces); (ii) differentiable actions of Lie groups on differentiable manifolds. As usual, there are the *local theory* as well as the *global theory* in both the *topological* and *differentiable* settings. In the *local theory*, one studies the geometric behavior of a topological (resp. differentiable) action in an invariant neighborhood of a *given orbit*. In the case of *compact differentiable* transformation groups, the differentiable slice theorem [cf. Th. (I.5)] shows that the local theory can be completely reduced to that of the *slice representation* of G_x on normal vectors. This *local linearity* is exactly the major reason that makes *compact differentiable* transformation groups much more regular and technically more accessible than either the *non-compact* case *or* the *topological* case. In the case of *topological* actions of compact Lie groups, one still has a *topological slice theorem* [cf. Th. (I.5′)]. However, the *local linearity* fail miserably in the *topological* case. This failure makes the local theory of topological compact transformation groups much more complicated and challenging. In § 1, we shall interprete the results of Chapter IV (specialized to the case of cohomology spheres and acyclic manifolds) as a kind of *splitting principle* and then proceed to introduce a geometric generalization of *weight system* which, for many purpose, is a *workable substitute of local linearity* for the study of *topological compact transformation groups.* The results of this chapter are mainly taken from a paper of the author [H 16].

§ 1. The Splitting Principle and the Geometric Weight System for Actions on Acyclic Cohomology Manifolds

Before embarking on a systematic investigation of the cohomological behavior of topological actions of compact connected Lie groups on acyclic cohomology manifolds, it is natural that we should first reflect philosophically and analyze technically the classical linear representation theory of compact connected Lie groups, so beautifully and profoundly accomplished by I. Schur, É. Cartan and H. Weyl. Basically, there are the following two crucial facts which constitute the foundation of the whole theory of linear representations of compact connected Lie groups, namely, (i) the *splitting* of complex linear representation of torus groups into direct sums of one-dimensional representations; (ii) the maximal tori theorem of É. Cartan and the Schur orthogonality relations (of representation functions) which enable us to reduce the classification of linear representation of compact connected Lie groups to that of their maximal tori. If one tries to put our successful experience in the *linear* case into a broader perspective of topological transformation groups, the crucial step is then to replace the *linear-splitting* of torus actions by some kind of workable splitting for topological torus actions. It is interesting to note that the so called *splitting principle* in the characteristic class theory of vector bundles exactly exploits the above *linear splitting* to obtain a useful *splitting in characteristic classes.* In this section, we shall show that, although a splitting at the *geometric level* for *topological* torus actions on acyclic manifolds is clearly out of the question, a splitting at the characteristic class level can actually be proved.

(A) *Equivariant Bundles and Equivariant Characteristic Classes*

Definition. An equivariant bundle is a fibre bundle $X \xrightarrow{p} Z$ such that both the total space X and the base space Z are *G-spaces* and the projection map p is *equivariant.*

Applying the Borel construction to a given equivariant bundle $X \xrightarrow{p} Z$, one obtains an "associated" bundle $X_G \xrightarrow{p_G} Z_G$. It is then rather natural to consider the characteristic classes of the bundle $X_G \xrightarrow{p_G} Z_G$ as the *equivariant* characteristic classes of the equivariant bundle $X \xrightarrow{p} Z$. For example, if $X \xrightarrow{p} Z$ is an equivariant vector bundle, then the Euler class (resp. the Pontrjgin classes) of $X_G \xrightarrow{p_G} Z_G$ are considered as the equivariant Euler class (resp. equivariant Pontrjgin classes) of $X \xrightarrow{p} Z$.

Example. Suppose G is a torus group and $\mathbb{R}^n = \mathbb{R}^m \oplus \mathbb{R}^r$ is a *linear* G-space with $\mathbb{R}^r = F$ as the fixed point set and

$$\Omega_0 = \{\pm w_1, k_1; \dots; \pm w_s, k_s\}$$

as the non-zero weight system. Then $\mathbb{R}^n \xrightarrow{p} \mathbb{R}^r = F$ is the *equivariant normal bundle* of the fixed point set F in $\mathbb{R}^n$, and it is not difficult to see that its *equivariant Euler class* is exactly the product of non-zero weights, namely, $\chi_G = (w_1^{k_1} \cdots w_s^{k_s}) \in H^m(B_G) \subseteq H^m(B_G \times F)$. Cohomologically, the above equivariant Euler class can also be characterized as follows:

Let S^{m-1} be the unit sphere of $\mathbb{R}^m$. Then it is obvious that the inclusion $S^{(m-1)} \subseteq \mathbb{R}^n - F$ induces isomorphism in cohomology. Hence, the E_2-term of the Serre spectral sequence of the fiberation

$$(\mathbb{R}^n - F) \to (\mathbb{R}^n - F)_G \to B_G$$

consists of only two lines, namely,

$$E_2^{p,q} = H^p(B_G) \otimes H^q(\mathbb{R}^n - F) \neq 0 \quad \text{only when} \quad q = 0 \qquad (m-1)$$

Moreover, the transgression of the generator $x \in H^{(m-1)}(\mathbb{R}^m - F) \cong H^{m-1}(S^{m-1})$ is exactly the above equivariant Euler class $\chi_G = (w_1^{k_1} \cdots w_s^{k_s}) \in H^m(B_G)$. Hence $\chi_G = (w_1^{k_1} \cdots w_s^{k_s})$ *is also the generator of the annahilator ideal of* $H_G^*(\mathbb{R}^n - F)$.

(B) *Equivariant Euler Class and the Geometric Weight System*

Let X be an acyclic cohomology manifold of dimension n with a given action of torus group G. Then the fixed point set F is also an acyclic cohomology submanifold, say, of dimension r, and $(X-F)$ is of cohomology type of $S^{(n-r-1)}$. Hence, it follows from the same spectral sequence argument that the annahilator ideal, A, of $H_G^*(X-F)$ is a principal ideal generated by the *transgression*, χ_G, of a generator $x \in H^{(n-r-1)}(X-F)$. Then, by Theorem (IV.6), the above generalized equivariant Euler class χ_G also *splits* into the product of linear polynomials, namely

$$\chi_G = w_1^{k_1} \cdots w_s^{k_s}, \qquad w_i \in H^2(B_G).$$

Remark. Suppose X is, to begin with, an acyclic cohomology manifold over $\mathbb{Z}$ (instead of over $\mathbb{Q}$). Then all the above $\mathbb{Q}$-cohomology algebras can be replaced by $\mathbb{Z}$-cohomology algebras, and finally, χ_G is an integral class which, by the Gauss lemma, also splits into the product of integral linear forms, i.e.,

$$\chi_G = c \cdot w_1^{k_1} \cdots w_s^{k_s}, \qquad w_i \in H^2(B_G; \mathbb{Z}).$$

Definition. Let G be a torus group and X be an acyclic cohomology manifold of dimension n with a given G-action. Then the annahilator ideal of $H_G^*(X-F)$ is principal and its generator, χ_G, splits into the product of linear forms, i.e., $\chi_G = w_1^{k_1} \cdots w_s^{k_s}$. The above system of weights $\pm w_i$ with multiplicity k_i $(i=1,\dots,s)$, is called the *system of non-zero weights of the G-space X*, namely, $\Omega'(X) = \{\pm w_1, k_1, \dots, \pm w_s, k_s\}$, and the *geometric weight system* of X, $\Omega(X)$, is defined to be $\Omega'(X)$ plus the zero weight with multiplicity $r = \dim F$.

The following theorem explains the geometric significance of the above weight system:

Theorem (V.I) [H 16]. *Let G be a torus and X be an acyclic cohomology manifold of dimension n with a given G-action. Suppose the geometric weight system of X is as follows:*

$$\Omega(X) = \{\pm w_1, k_1; \dots; \pm w_s, k_s; 0, r\}.$$

Then (i) *a subgroup $K \in \mathcal{O}^0(X)$ if and only if K is of the form*

$$K = w_{i_1}^{\perp} \cap \cdots \cap w_{i_k}^{\perp} \qquad (K = G \text{ if } k=0),$$

(ii) *the F^0-variety of x, $F^0(x)$, is a acyclic cohomology manifold whose dimension is given by*

$$\dim F^0(x) = r + \textstyle\sum_{w_i^{\perp} \supseteq G_x^0} 2k_i.$$

Remarks. (i) The above theorem proves that the connected orbit types of $X, \mathcal{O}^0(X)$, and the connected F-variety structure of X are *cohomologically* the same as that of the linear example with the same weight system, i.e., $\Omega(\varphi) = \Omega(X)$.

(ii) Since we use rational coefficients in the above theorem, we can only detect the connected isotropy subgroups. However, if one uses integral coefficients, it is possible to obtain some information about (G_x/G_x^0). For example, in a recent paper of Golber [G 3], he gave a neat geometric meaning to the integral factor c in the factorization of $\chi_G = c \cdot w_1^{k_1} \cdots w_s^{k_s}$.

(iii) The above theorem is essentially an improvement of the Borel formula which relates the dimension of F^0-varieties of corank $\leqslant 1$ to the dimension of X.

Proof. Let $K \subseteq G$ be a subtorus and P_K be the kernel of $H^*(B_G) \to H^*(B_K)$. Then $X^K = F(K, X)$ is an invariant acyclic cohomology submanifold of X; and it follows easily from the localization theorem that

$$\operatorname{ann}(H_G^*(X-F))_{P_K} = \operatorname{ann}(H_G^*(X^K - F))_{P_K}.$$

Hence, the ideal $\operatorname{ann}(H_G^*(X^K - F))$ is generated by

$$\chi_G(X^K) = \prod_{w_i^\perp \supseteq K} w_i^{k_i}$$

and

$$\dim X^K = r + \sum_{w_i^\perp \supseteq K} 2k_i \,.$$

Notice that a top dimensional submanifold of a cohomology manifold is automatically open and X^K are always closed. Hence, it follows easily from the connectedness of X^K (which is always acyclic) that $X^K = X^{K'}$ if and only if

$$w_i^\perp \supseteq K \Leftrightarrow w_i^\perp \supseteq K' \quad \text{for all} \quad 1 \leqslant i \leqslant s\,,$$

and the two assertions of the above theorem follows immediately. □

Definition. Let G be a compact connected Lie group and T be a maximal torus of G. Let X be an acyclic cohomology manifold with a given G-action. Then the weight system of the G-space X is simply defined to be the weight system of the restricted T-space X.

(C) *System of Local Weights*

Next let us consider the *local* situation around a fixed point x in a cohomology G-manifold X. It is not difficult to show that the family of *invariant* neighborhoods of x forms a *cofinal* subset in the family of *all* neighborhoods of x. Hence, in a limiting sense, $\lim H^*(U-x) \cong H^*(S^{n-1})$ for the directed system of invariant neighborhoods U of x. It is rather straightforward to modify the above discussion so that one may define a system of *local* weights at the fixed point x as if there is actually an acyclic invariant neighborhood U of x. Furthermore, one also has the local version of Theorem (V.1) which explains the geometric meaning of the system of local weights.

Theorem (V.I′). *Let G be a torus group, X be a cohomology G-manifold of dimension n and x be a fixed point in X. Suppose that the system of local weights at x is as follows:*

$$\Omega_x(X) = \{\pm w_1, k_1; \ldots; \pm w_s, k_s; 0, r\}\,.$$

Then (i) *K is the generic connected isotropy subgroup of an F^0-variety Y passing through x if and only if K is of the form*

$$K = w_{i_1}^\perp \cap \cdots \cap w_{i_k}^\perp \quad (K = G \text{ if } k=0)\,,$$

(ii) *the dimension of an F^0-variety Y passing through x is given by*

$$\dim Y = r + \sum_{w_i^\perp \supseteq G_Y} 2k_i \,.$$

Corollary 1. *Suppose X is a cohomology G-manifold and x, y are two fixed points in X. If x, y lie in the same connected component of the fixed point at $F = F(G, X)$, then the systems of local weights of x and y are the same, i.e., $\Omega_x(X) = \Omega_y(X)$.*

Proof. Since x, y belong to the same connected component of the fixed point set, it follows from the definition of F^0-variety that the set of F^0-varieties passing through x is identically the same as the set of F^0-varieties passing through y. Hence, it follows from the above theorem that $\Omega_x(X) = \Omega_y(X)$. □

Corollary 2. *The weight system $\Omega_x(X)$ (resp. $\Omega(X)$, if X is itself acyclic) is invariant under the action of $W(G)$.*

Proof. Let us show the case X is an acyclic manifold. Then $F(T, X)$ is also acyclic and hence connected. Notice that $F(T, X)$ is clearly invariant under $W(G)$ and $\sigma\Omega_x(X) = \Omega_{\sigma(x)}(X)$ for $\sigma \in W(G)$ and $x \in F(T, X)$. Hence Corollary 2 follows directly from Corollary 1 and the connectedness of $F(T, X)$. □

§ 2. Geometric Weight System and Orbit Structure

In this section, we shall always assume that G is a compact connected Lie group, T is a maximal torus of G and X is an acyclic cohomology G-manifold. In the last section, a simple invariant called the *geometric weight system* of X was defined in terms of the restricted T-action on X. Moreover, Theorem (V.1) demonstrates that the cohomological aspects of the orbit structure of the *restricted T-action* can be read off from the weight system $\Omega(X)$. Hence, it is natural to investigate how much of the *orbit structure of the G-action* itself can be determined by the weight system $\Omega(X)$.

(A) *The Topological Slice Theorem and Weight System*

The topological slice theorem [cf. § 3, Ch. I] asserts that, to a given point x of a G-space X, there exists a slice, S_x, satisfying the following two properties: (i) S_x is invariant under G_x, (ii) $G(S_x) = G \times_{G_x} S_x$ is an invariant neighborhood of the orbit $G(x)$. In the special case that X is a cohomology manifold, then it is obvious that S_x is also a cohomology manifold with $\dim(S_x) = (\dim X - \dim G(x))$.

Definition. The weight system of the slice at x, denoted by $\Omega(S_x)$, is defined to be the *system of local weights at x in the G_x-space S_x*, namely, $\Omega(S_x) = \Omega_x(S_x)$.

Proposition 1. *Let G_1, G_2 be compact connected Lie groups, $G_1 \xrightarrow{h} G_2$ be a Lie group homomorphism and T_1, T_2 be maximal tori of G_1, G_2 respectively such that $h(T_1) \subseteq T_2$. Suppose X is a given acyclic cohomology G_2-manifold and $h^*(X)$ be the G_1-space structure on X induced by h. Then*

$$\Omega(h^*(X)) = h^* \Omega(X) = \{h^* w_i, k_i\}$$

where $h^* w_i$ is the image of w_i under the induced map $h^*: H^2(B_{T_2}) \to H^2(B_{T_1})$. In particular, if G_1 is a subgroup of G_2 and $h: G_1 \to G_2$ is the inclusion homomorphism, we shall simply denote $\Omega(h^*(X)) = h^* \Omega(X)$ by $\Omega(X)|G_1$, or by $\Omega(X)|T_1$.

Proof. Straightforward verification.

Proposition 2. *Let X be a given acyclic cohomology G-manifold with weight system $\Omega(X)$. For a given point $x \in X$, one may assume that the maximal torus T_1 of G_x^0 is contained in the maximal torus T of G. Then there exist a suitable subcollection of weights in $\Omega(X)$ such that*

$$T_1 = G_x^0 \cap T = T_x^0 = w_{j_1}^{\perp} \cap \cdots \cap w_{j_t}^{\perp}$$

and the weight system of the slice at x, $\Omega(S_x)$, is given by

$$\Omega(X)|T_1 = \Omega(\mathrm{Ad}_G|T_1 - \mathrm{Ad}_{G_x}|T_1) + \Omega(S_x).$$

Proof. Since $T_1 = T_x^0$, it follows directly from Theorem (V.1) that there exist suitable weights $\{w_{j_1}, \ldots, w_{j_t}\}$ in $\Omega(X)$ such that $T_1 = w_{j_1}^{\perp} \cap \cdots \cap w_{j_t}^{\perp}$. Let q be a fixed point of the maximal torus T of G. Then, it is clear that $\Omega(X)|T_1$ is the system of local weights at q in the T_1-space X and

$$\Omega(\mathrm{Ad}_G|T_1 - \mathrm{Ad}_{G_x}|T_1) + \Omega(S_x)$$

is the system of local weights at x in the T_1-space X. Hence, the equality follows from the connectedness of $F(T_1, X)$, which is in fact acyclic, and from Corollary 1 of Theorem (V.1). □

Corollary 1. *Let (H) be the principal orbit type of the acyclic cohomology G-manifold X. Then the above equation reduces to the following simple equation*

$$\Omega(X)|H \equiv \Omega(\mathrm{Ad}_G|H - \mathrm{Ad}_H) \quad (\text{mod } \textit{zero weights}).$$

Corollary 2. *Let $\Delta(G)$ and $\Delta(G_x)$ be the root systems of G and G_x respectively, $x \in F(T, X)$. Then*

$$\Delta(G_x) \supseteq \Delta(G) \setminus \Omega(X) \quad (\textit{the difference set}).$$

Definition. A subgroup $H \subseteq G_1 \times G_2 \times \cdots \times G_k$ is called a *splitting subgroup* (with respect to the given decomposition) if

$$H = (H \cap G_1) \times (H \cap G_2) \times \cdots \times (H \cap G_k).$$

Definition. The weight system $\Omega(X)$ of an acyclic cohomology G-manifold X *is called splitting* with respect to the decomposition of $G = G_1 \times G_2 \times \cdots \times G_k$ if

$$\Omega(X) = \Omega(X)|G_1 + \Omega(X)|G_2 + \cdots + \Omega(X)|G_k \quad (\text{mod zero weights}).$$

Proposition 3. *If the weight system $\Omega(X)$ of an $\mathbb{Q}$-acyclic cohomology G-manifold X is splitting with respect to the decomposition of $G = G_1 \times G_2 \times \cdots \times G_k$, then the connected isotropy subgroups G_x^0 are splitting subgroups for all $x \in X$.*

Proof. Let $T, T_1, \ldots, T_k$ be maximal tori of $G, G_1, \ldots, G_k$ respectively, and $T = T_1 \times \cdots \times T_k$. Consider the restricted T-action on X. Since $\Omega(X)$ is splitting, it follows from (i) of Theorem (V.1) that $\{T_x^0\}$ are splitting subgroups for all $x \in X$. Then, the proposition follows from the following lemma of [H14]. □

Lemma. *A connected subgroup $H \subseteq G = G_1 \times \cdots \times G_k$ is a splitting subgroup if and only if all its maximal tori $T' \subseteq H \subseteq G$ are also splitting subgroups.*

Proof. Let $h \in H$ be an arbitrary element of H and π_j be the projection of G onto its j-th factor G_j. Then, it is easy to see that H is a splitting subgroup if and only if $\pi_j(h)$ also belongs to H for $j = 1, \ldots, k$. Let T' be a maximal torus of H containing h. Since T' is assumed to be splitting, $\pi_j(h) \in T' \subseteq H$. □

Conjecture. In case $G = G_1 \times \cdots \times G_k$ is connected, X is a $\mathbb{Z}$-acyclic cohomology manifold and $\Omega(X)$ is splitting, the isotropy subgroups G_x themselves should also be splitting subgroups. [In the differentiable setting, the above conjecture was proved in [H 14].]

(B) *Actions of Classical Groups with Simple Weight Patterns*

Let $G = \mathrm{SO}(n)$ (resp. $\mathrm{SU}(n), \mathrm{Sp}(n)$) be a classical group and ρ_n (resp. μ_n, ν_n) be the standard linear action of G on $\mathbb{R}^n$ (resp. $\mathbb{C}^n, \mathbb{H}^n$). We shall study the orbit structures of the acyclic cohomology G-manifolds X with weight systems, $\Omega(X)$, modelled after several copies of such standard linear actions, namely,

$$\Omega(X) \equiv (l \cdot \rho_n) \text{ (resp. } \Omega(l \cdot \mu_n), \Omega(l \cdot \nu_n)) \quad \text{(mod zero weights)}.$$

Let $(\theta_1, \theta_2, \ldots, \theta_m)$ be the usual coordinates of a maximal torus T of G, where

$$m = \begin{cases} \left[\dfrac{n}{2}\right] & \text{if } G = \mathrm{SO}(n) \\ n & \text{and } \theta_1 + \cdots + \theta_n = 0 \text{ if } G = \mathrm{SU}(n) \\ n & \text{if } G = \mathrm{Sp}(n)\,. \end{cases}$$

Then

$$\Omega(X) \equiv \left.\begin{cases} \{\pm\theta_i, l\} & \text{if } G = \mathrm{SO}(n), \mathrm{SU}(n) \\ \{\pm\theta_i, 2l\} & \text{if } G = Sp(n) \end{cases}\right\} \text{(mod zero weights)}.$$

Theorem (V.2). *Let $G = \mathrm{SO}(n)$ (resp. $\mathrm{SU}(n), \mathrm{Sp}(n)$) and X be an acyclic cohomology G-manifold with the system of non-zero weights $\Omega'(X) = \{\pm\theta_i, l\}$. Then*

(i) *all connected isotropy subgroups G_x^0 are conjugate to the standard $\mathrm{SO}(k)$ (resp $\mathrm{SU}(k), \mathrm{Sp}(k)$) for suitable $k \leq n$,*

(ii) *the connected principal isotropy subgroup type (H) is non-trivial when and only when $l \leq (n-2)$ (resp. $l \leq (n-2)$, $l \leq 2(n-1)$) and $H = \mathrm{SO}(n-l)$ (resp. $\mathrm{SU}(n-l)$, $Sp(n-l/2)$); l must be even when $G = Sp(n)$ and $l < 2n$).*

Proof. Let $x \in X$ be an arbitrary point of X. Up to conjugation, we may assume $(G_x \cap T)^0$ is a maximal torus of G_x^0. It follows from (i) of Theorem (V.1) that

$$T_1 = T_x^0 = (G_x \cap T)^0 = \theta_{j_1}^{\perp} \cap \cdots \cap \theta_{j_t}^{\perp}.$$

Since there exists an element of the Weyl group $W(G)$ which maps $\theta_{j_1},\dots,\theta_{j_t}$ to $\theta_1,\dots,\theta_t$ respectively, we may assume the maximal torus of G_x^0 to be $\theta_1^\perp\cap\cdots\cap\theta_t^\perp$. Now, it follows from Corollary 2 of Proposition 2 that

$$\Delta(G_x^0)\supseteq\Delta(G)|T_1\setminus\Omega(X)|T_1\,.$$

Then, it is an easy exercise in Lie algebra theory to show

$$G_x^0=\mathrm{SO}(k),\qquad \left[\frac{k}{2}\right]=(m-t)\qquad (\text{resp. } \mathrm{SU}(n-t),\ \mathrm{Sp}(n-t))\,.$$

Similarly, it follows from Corollary 1 of Proposition 2 that the connected principal isotropy subgroups $(H)=(\mathrm{SO}(n-l))$ (resp. $\mathrm{SU}(n-l), \mathrm{Sp}(n-(l/2))$; l even if $G=\mathrm{Sp}(n)$ and $l<2n$). □

Remark. The above theorem is the "topological version" of Theorem 2 of [H 14] for differentiable actions of classical groups on acyclic manifolds. However, due to the fact that topological weight system only retains the direction of the weight vectors, the above theorem only determines the connected component of the isotropy subgroups. In fact, it is possible to prove that G_x are actually connected if one similarly introduce the concept of p-weight system $\Omega_p(X)$ and also assume that $\Omega'_p(X)=\Omega'_p(l\cdot\rho_n)$ (resp. $\Omega'_p(l\cdot\mu_n), \Omega'_p(l\cdot\nu_n)$).

Next we consider the following two cases:

(i) $G=\mathrm{SU}(n),\ \Omega'(X)=\Omega'(\Lambda^2\mu_n)=\{\pm(\theta_i+\theta_j);\ i<j\}$,

(ii) $G=\mathrm{Sp}(n),\ \Omega'(X)=\Omega'(\Lambda^2\nu_n)=\{\pm\theta_i\pm\theta_j;\ i<j\}$.

Theorem (V.3). *Let $G=\mathrm{SU}(n)$ (resp. $\mathrm{Sp}(n)$) and X be an acyclic cohomology G-manifold. If $\Omega'(X)=\Omega'(\Lambda^2\mu_n)$ (resp. $\Omega'(\Lambda^2\nu_n)$) then the connected orbit types of X, $\mathcal{O}^0(X)$, is the same as that of $\Lambda^2\mu_n$ (resp. $\Lambda^2\nu_n$). In particular, the connected principal orbit type $(H)=(\mathrm{SU}(2)^{[n/2]})$ (resp. $(\mathrm{Sp}(1)^n)$). Moreover, $F(G_x^0,X)$ is also an acyclic cohomology manifold and $\operatorname{codim}F(G_x^0,X)=\operatorname{codim}F(G_x^0,V)$ in the representation space V of $\Lambda^2\mu_n$ (resp. $\Lambda^2\nu_n$).*

Theorem (V.4). *Let G be a compact connected Lie group and X be an acyclic cohomology G-manifold with $\Omega'(X)=\Delta(G)$. Then* (i) *the principal orbit type is the conjugacy class of maximal tori,*

(ii) *the Weyl group $W(G)$ acts as a group generated by (topological) reflections on $Y=F(T,X)$,*

(iii) *the natural map $Y/W\to X/G$ induced by the inclusion $Y\subseteq X$ is a bijection.*

Proof. We refer to [H 16] for the proof of the above two theorems.

(C) *The Determinantion of Principal Orbit Type in Terms of $\Omega(X)$*

Let G be a compact connected Lie group and X be an acyclic cohomology G-manifold. Since the principal orbit type (H_X) of X is a dominant geometric characteristic of the G-manifold X, it is natural to try to determine the principal orbit type (H_X) of X from the weight system, $\Omega(X)$, of X. The main result of the subsection is the following algorithm which enables us to compute the *connected principal orbit type* (H_X^0) from the system of non-zero weights $\Omega'(X)$.

Theorem (V.5). *Let $T \subseteq G$ be the maximal torus that one uses to define the weight system $\Omega(X)$, and (H_X^0) be the connected principal orbit type of the G-manifold X. Up to a conjugation, one may assume that $S=(H_X \cap T)^0$ is a maximal torus of H_X^0. Then there exist a sequence of non-zero weights $w_1, \ldots, w_k \in \Omega'(X)$ together with a sequence of decreasing subtori $T=S_0 \supsetneqq S_1 \supsetneqq \cdots \supsetneqq S_k=S$ which satisfy the following recursive conditions:*

$$S_0=T,\ w_1 \in \{\Omega'(X)-\Omega'(\mathrm{Ad}_G)\} \neq \emptyset$$
$$S_1=w_1^{\perp},\ w_2 \in \{\Omega'(X)|S_1-\Omega'(\mathrm{Ad}_G|S_1)\} \neq \emptyset$$
- -
$$S_i=S_{i-1} \cap w_i^{\perp},\ w_{i+1} \in \{\Omega'(X)|S_i-\Omega'(\mathrm{Ad}_G|S_i)\} \neq \emptyset$$
- -
$$S_k=S_{k-1} \cap w_k^{\perp},\ \{\Omega'(X)|S_k-\Omega'(\mathrm{Ad}_G|S_k)=\emptyset \quad (empty).$$

Conversely, suppose there exist a sequence of non-zero weights and a sequence of decreasing subtori from T to S satisfying the above recursive conditions. Then S is a maximal torus of a suitable connected principal isotropy subgroup H_X^0 and the root system, $\Delta(H_X^0)$, of H_X^0 is given by the following equation

$$\Delta(H_X^0)=\{\Omega'(\mathrm{Ad}_G|S_k)-\Omega'(X)|S_k\}.$$

Remark. If $\{\Omega'(X)-\Omega'(\mathrm{Ad}_G)\}=\emptyset$ (empty), then $k=0$ and $S_k=T$. The difference $\{\Omega'(X)|S_i-\Omega'(\mathrm{Ad}_G|S_i)\}$ is a difference of sets *with multiplicities*, namely, the multiplicity of a given weight w in the difference set is equal to the difference of multiplicities if it is positive and zero otherwise.

The above theorem has the following useful corollaries:

Corollary 1. *Let X_1, X_2 be two acyclic cohomology G-manifolds with the same system of non-zero weights, i.e., $\Omega'(X_1)=\Omega'(X_2)$. Then they also have the same connected principal orbit type, i.e., $(H_{X_1}^0)=(H_{X_2}^0)$.*

Corollary 2. *Let X, X_1, X_2 be acyclic cohomology G-manifolds. If $\Omega'(X)=\Omega'(X_1)+\Omega'(X_2)$ (sum of sets with multiplicity), then (H_X^0) is the intersection in general position of $(H_{X_1}^0)$ and $(H_{X_2}^0)$, namely, (H_X^0) is the conjugacy class of those smallest possible intersections among $\{g_1 H_{X_1}^0 g_1^{-1} \cap g_2 H_{X_2}^0 g_2^{-1}\}$.*

Proof of Corollary 2. Let $X_1 \times X_2$ be the product G-space of X_1, X_2, i.e., $g \cdot (x_1, x_2)$ for all $x_1 \in X_1$ and $x_2 \in X_2$. Then it is clear that $\Omega'(X_1 \times X_2)=\Omega'(X_1)+\Omega'(X_2)$ and $(H_{X_1 \times X_2}^0)$ is the intersection in general position of $(H_{X_1}^0)$ and $(H_{X_2}^0)$. Hence Corollary 2 follows from Corollary 1 and the assumption $\Omega'(X)=\Omega'(X_1)+\Omega'(X_2)=\Omega'(X_1 \times X_2)$. □

Corollary 3. *Let X_1, X_2 be acyclic cohomology G-manifolds and*

$$\Omega'(X_1) \supseteq \Omega'(X_2) \quad (as\ sets\ with\ multiplicity).$$

Then $(H^0_{X_1}) \leqslant (H^0_{X_2})$ *in the sence that* $H^0_{X_1}$ *is conjugate to a subgroup of* $H^0_{X_2}$.

Proof of Corollary 3. Up to conjugation, we may assume that $(H^0_{X_2} \cap T) = S''$ is a maximal torus of $H^0_{X_2}$. By Theorem (V.5), there exist a sequence of weights and a decreasing sequence of subtori from T to S'' satisfying the recursive conditions:

$$S_i = S_{i-1} \cap w_i^{\perp}; \qquad w_{i+1} \in \{\Omega'(X_2)|S_i - \Omega'(\mathrm{Ad}_G|S_i)\}.$$

Since we assume that $\Omega'(X_1) \supseteq \Omega'(X_2)$, it is clear that

$$\{\Omega'(X_1)|S - \Omega'(\mathrm{Ad}_G|S)\} \supseteq \{\Omega'(X_2)|S - \Omega'(\mathrm{Ad}_G|S)\}$$

holds for any subtorus $S \subseteq T$. Hence, if $\{\Omega'(X_1)|S'' - \Omega'(\mathrm{Ad}_G|S'')\}$ is also empty, then S'' is also a maximal torus of a suitable $H^0_{X_1}$. Otherwise, we may continue the sequences for $\Omega'(X_2)$

$$\{w_1, w_2, \ldots, w_{k''}\}, \qquad \{T = S_0 \supsetneqq S_1 \supsetneqq \cdots \supsetneqq S_{k''} = S''\}$$

to obtain the corresponding sequences for $\Omega'(X_1)$, i.e.,

$$\{w_1, \ldots, w_{k''}, \ldots, w_{k'}\}, \qquad \{T = S_0 \supsetneqq \cdots \supsetneqq S_{k''} \supsetneqq \cdots \supsetneqq S_{k'} = S'\}$$

so that S' is a maximal torus of $H^0_{X_1}$. Notice that $S' \subseteq S'$ and

$$\Delta(H^0_{X_1}) = \{\Omega'(\mathrm{Ad}_G|S') - \Omega'(X_1)|S'\} \subseteq \{\Omega'(\mathrm{Ad}_G|S') - \Omega'(X_2)|S'\} = \Delta(H^0_{X_2})|S'$$

which clearly implies that $(H^0_{X_1}) \leqslant (H^0_{X_2})$.

We shall need the following two lemmas in the proof of Theorem (V.5).

Lemma 1. *If* $\{\Omega'(X) - \Delta(G)\} = \emptyset$ *(empty), then the connected principal isotropy subgroups* (H^0_X) *are of maximal rank.*

Since $\Omega'(X) \subseteq \Delta(G)$ and $\Delta(G)$ is completely splitting (i.e., splitting with respect to any decomposition of G), it is clear that $\Omega'(X)$ is also completely splitting and it is easy to apply Proposition 3 to reduce the proof of Lemma 1 to the special case that G is *simple*. However, in case G is simple, one has the following more precise result:

Lemma 1'. *Let* G *be a simple compact connected Lie group and* X *be an acyclic cohomology* G*-manifold with* $\Omega'(X) \subseteq \Delta(G)$. *Then either*

(i) $\Omega'(X) = \Delta(G)$ *and* $(H^0_X) = (T)$ *[maximal tori],*

or (ii) $\Omega'(X) = \{$*the set of short roots of* $G\}$ *and respectively*

$$(H^0_X) = \begin{Bmatrix} (D_n) \\ (Sp(1)^n) \\ (\mathrm{Spin}(8)) \\ (\mathrm{SU}(3)) \end{Bmatrix} \quad \textit{where} \quad G = \begin{cases} B_n \\ C_n \\ F_4 \\ G_2. \end{cases}$$

Proof of Lemma 1'. The case $\Omega'(X)=\Delta(G)$ is covered by Theorem (V.4), we shall only consider the case $\phi\neq\Omega'(X)\subsetneq\Delta(G)$. Since $\Omega'(X)$ and $\Delta(G)$ are both invariant under the Weyl group $W(G)$, and it is a well known fact that the root system $\Delta(G)$ of a *simple* compact connected Lie group G consists of at most two orbits of $W(G)$, the above case is possible only when $\Delta(G)$ consists of *two orbits* of $W(G)$ and $\Omega'(X)$ is equal to one of them. Hence $G=B_n, C_n, F_4$ or G_2 and

$$\Delta(G)=\{\text{the long roots}\}+\{\text{the short roots}\}.$$

Let $\alpha\in\Omega'(X)\subseteq\Delta(G)$ be a root in $\Omega'(X)$, and G_α be the connected normalizor of $T_\alpha=\alpha^\perp$, $\bar{G}_\alpha=G_\alpha/T_\alpha$. Let $Y_\alpha=F(T_\alpha,X)$. Then the induce action of $\bar{G}_\alpha$ on Y_α has the weight system $\Omega'(Y_\alpha)=\Delta(\bar{G}_\alpha)=\{\pm\alpha\}$. This is the simplest and the most basic situation considered in Theorem (V.4). It is not difficult to prove that [cf. Lemma 1, § 5, Ch. V]

$$F(G_\alpha,X)=F(\bar{G}_\alpha,Y_\alpha)=F(\sigma_\alpha,F(T,X))$$

is a codimension one submanifold of $F(T,X)$, i.e., the induced action of the order 2 generator $\sigma_\alpha\in W(G)$ on $F(T,X)$ is a topological reflection. Hence, there exists $y\in F(T,X)$ with $T\subseteq G_y\subsetneq G$. On the other hand, we claim that $\Delta(G)-\Delta(K)$ consists of at least one short root for any *proper maximal rank* subgroup $K\subsetneq G$. Of course, we need only to check the above assertion for those maximal, maximal rank subgroups of G (see [B12] for detail account of such subgroups). Since the Dynkin diagram of such subgroups K are *subdiagrams of the diagram of Cartan polyhedon of* G, namely

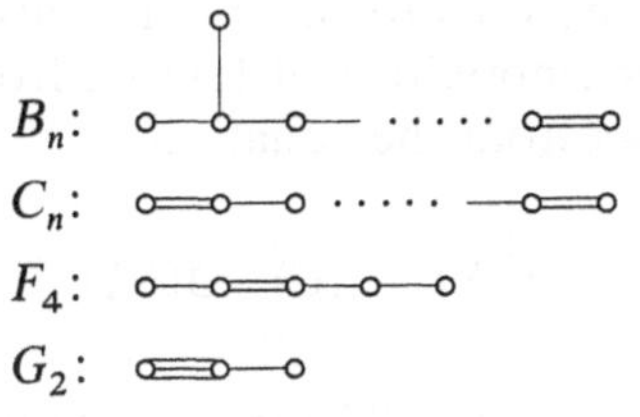

it is not difficult to check that $\Delta(K)$ *does not contain all* the short roots of G. Hence, it follows from Proposition 2 that

$$\Omega'(X)=\{\Delta(G)-\Delta(G_y)\}+\Omega'(S_y)$$

and consequently $\Omega'(X)$ contains some short roots. Therefore, $\Omega'(X)$ = {the set of short roots}, and the above lemma follows easily from a straightforward computation using the equation of Corollary 1 of Proposition 2. □

Lemma 2. *Suppose that* $\{\Omega'(X)-\Delta(G)\}$ *is non-empty. Then to any weight* $w\in\{\Omega'(X)-\Delta(G)\}$, *there exists a point* $x\in X$ *such that* $T_x^0=w^\perp=G_x^0\cap T$ *and is a maximal torus of* G_x^0.

Proof of Lemma 2. Let $K=G_y$, $y\in F(T,X)$ be *minimal* among the set of all *maximal rank* isotropy subgroups, and S_y be the slice at x which is a K-space.

Then

$$\Omega'(X)=\{\Delta(G)-\Delta(K)\}+\Omega'(S_y) \Rightarrow \{\Omega'(X)-\Delta(G)\}=\{\Omega'(S_y)-\Delta(K)\}.$$

On the other hand, it follows from Theorem (V.1), applying to the K-space S_y, that to any $w\in\{\Omega'(S_y)-\Delta(K)\}=\{\Omega'(X)-\Delta(G)\}$ there exists a point $x\in S_y$ such that $T_x^0=w^\perp$. Since $K=G_y$ is chosen to be minimal among all maximal rank isotropy subgroups

$$\mathrm{rk}(G)-1=\mathrm{rk}(T_x)\leqslant \mathrm{rk}(G_x)<\mathrm{rk}(K)=\mathrm{rk}(G).$$

Hence $\mathrm{rk}(G_x)=(\mathrm{rk}(G)-1)=\mathrm{rk}(T_x^0)$ and $T_x^0=w^\perp$ is a maximal torus of G_x^0. □

Proof of Theorem (V.5). We shall prove Theorem (V.5) by induction on compact connected Lie groups.

(1) In the case $G=T$ is a torus, then $\Delta(G)=\emptyset$ and $H^0=\mathrm{Ker}^0(X)$ is the connected part of the ineffective kernel of the G-action on X. On the other hand, it follows from Theorem (V.1) that

$$H^0=G_x^0=\bigcap\{w^\perp, w\in\Omega'(X)\}.$$

Hence, Theorem (V.5) holds for the case $G=T$.

(2) Now, we assume that Theorem (V.5) holds for all proper compact connected subgroups of G and proceed to show that it also holds for G itself.

If $\{\Omega'(X)-\Delta(G)\}=\emptyset$ is empty, then it follows from Lemma 1 that (H_X^0) are maximal rank and it follows from the equation

$$\Omega'(X)=\Delta(G)-\Delta(H_X^0)$$

that $\Delta(H_X^0)=\{\Delta(G)-\Omega'(X)\}$. Hence Theorem (V.5) holds when $\{\Omega'(X)-\Delta(G)\}$ is empty.

Next, let us consider the case where $\{\Omega'(X)-\Delta(G)\}$ is non-empty. Then, by Lemma 2, there exists $y\in X$ such that $S_1=w_1^\perp$ is a maximal torus of G_y^0, where $w_1\in\{\Omega'(X)-\Delta(G)\}$. By Proposition 2, we have

$$\Omega'(X)|S_1=\{\Omega'(\mathrm{Ad}_G|S_1)-\Delta(G_y)\}+\Omega'(S_y)$$

therefore,

$$\{\Omega'(X)|S-\Omega'(\mathrm{Ad}_G|S)\}=\{\Omega'(S_y)|S-\Omega'(\mathrm{Ad}_{G_y}|S)\}$$

for any subtorus $S\subseteq S_1$. On the other hand, since the principal orbits are everywhere dense, it is clear that the principal isotropy subgroups of the G_y-space S_y are also principal isotropy subgroups of X. Hence, Theorem (V.5) follows by applying the induction assumption to the G_y-space S_y. □

§ 3. Classification of Principal Orbit Types for Actions of Simple Compact Lie Groups on Acyclic Cohomology Manifolds

In view of the principal orbit type theorem of Montgomery-Samelson-Yang [M 6], the principal orbits are population-wise dominating everywhere and hence the type of principal orbits is a geometric characteristic of fundamental importance. On the other hand, for spaces of a given specific type, the *possibilities of principal orbit types* are usually rather limited. Hence, in the study of topological actions of a given compact connected Lie group G on spaces of a certain type (such as acyclic cohomology manifolds as we do in this chapter), one of the natural problems of primary importance is the *classification of principal orbit types* for all topological G-actions on spaces of a given type. In this section, we shall first work out the classification of principal orbit types for *topological actions* of those *simple compact connected* Lie groups on acyclic manifolds.

Since the family of linear actions usually offers one of the most valuable classes of typical examples for such a purpose, it is natural to work out the much easier problem of classifying principal orbit types for *linear actions* of simple compact connected Lie groups.

(A) *Classification of Connected Principal Isotropy Subgroups for Linear Actions of Simple Compact Connected Lie Groups*

The classification of *linear actions* of *simple* compact connected Lie groups with non-trivial connected principal isotropy subgroups, i.e., $H_\psi^0 \neq \{\mathrm{id}\}$, has already been carried out in [H 5] and [K 4] independently. We list the result as follows:

Table A: *Real irreducible representations of simple compact connected Lie groups with non-trivial connected principal isotropy subgroups:*

Notations. We shall use the usual Lie algebra terminology. Let $\mathfrak{g}$ be a simple compact Lie algebra of rank r and $\Pi(G) = \{\alpha_1, \alpha_2, \ldots, \alpha_r\}$ be a system of primitive roots of $\mathfrak{g}$. We shall identify a real representation ψ with its complexification and denote the r basic representation corresponding to $\alpha_1, \alpha_2, \ldots, \alpha_r$ by $\phi_1, \phi_2, \ldots, \phi_r$ respectively.

I. $\mathfrak{g} = A_r$: $\alpha_1\ \alpha_2 \cdots \alpha_{r-1}\ \alpha_r$ (o-o- ··· -o—o)

rank r	ψ	$\Omega'(\psi)$	(H_ψ^0)
$r \geqslant 1$	Ad	$\Delta(\mathfrak{g})$	(T); maximal tori
$r \geqslant 2$	$\phi_1 + \phi_r$	$\{\pm\theta_i, i = 1, \ldots, r+1\}$	$(A_{(r-1)})$
$r \geqslant 4$	$\phi_2 + \phi_{r-1}$	$\{\pm(\theta_i + \theta_j), i<j\}$	$(SU(2)^{[(r+1)/2]})$
$r = 3$	ϕ_2	$\{(\theta_i + \theta_j), i<j\}$	$(B_2 = C_2)$
$r = 5$	$2\phi_3$	$2 \cdot \{(\pm\theta \pm \theta_j \pm \theta_k), i<j<k\}$	$T^2 \subseteq SU(3) \times SU(3) \subseteq SU(6)$

II. $\mathfrak{g}=B_r, r\geqslant 3$: 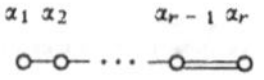

rank r	ψ	$\Omega'(\psi)$	(H_ψ^0)
$r\geqslant 3$	Ad	$\Delta(\mathfrak{g})$	(T); maximal tori
$r\geqslant 3$	ϕ_1	$\{\pm\theta_i, i=1,\dots,r\}$	(D_r)
$r=3$	ϕ_3	$\{\frac{1}{2}(\pm\theta_1\pm\theta_2\pm\theta_3)\}$	(G_2); $\dfrac{\text{Spin}(7)}{G_2}=S^7$
$r=4$	ϕ_4	$\{\frac{1}{2}(\pm\theta_1\pm\theta_2\pm\theta_3\pm\theta_4)\}$	$(\text{Spin}(7))$; $\dfrac{\text{Spin}(9)}{\text{Spin}(7)}=S^{15}$

III. $\mathfrak{g}=C_r, r\geqslant 2$: 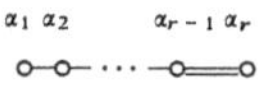

rank r	ψ	$\Omega(\psi)$	(H_ψ^0)
$r\geqslant 2$	Ad	$\Delta(\mathfrak{g})$	(T); maximal tori
$r\geqslant 2$	$2\phi_1$	$2\cdot\{\pm\theta_i, i=1,\dots,r\}$	(C_{r-1})
$r\geqslant 2$	ϕ_2	$\{\pm\theta_i\pm\theta_j, i<j\}$	$(\text{Sp}(1)^r)$

IV. $\mathfrak{g}=D_r, r\geqslant 4$: 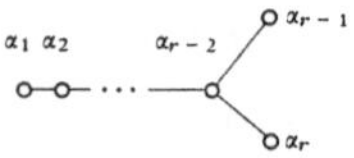

rank r	ψ	$\Omega'(\psi)$	(H_ψ^0)
$r\geqslant 4$	Ad	$\Delta(\mathfrak{g})$	(T); maximal tori
$r\geqslant 4$	ϕ_1	$\{\pm\theta_i, 1=1,\dots,r\}$	(B_{r-1})
$r=5$	$\phi_4+\phi_5$	$\{\frac{1}{2}(\pm\theta_1\pm\theta_2\pm\theta_3\pm\theta_4\pm\theta_5)\}$	$(\text{SU}(4))$
$r=6$	$2\phi_5$ or $2\phi_6$	$\{\frac{1}{2}(\pm\theta_1\pm\cdots\pm\theta_6)\}$with even or odd -1	$(\text{SU}(2)\times\text{SU}(2)\times\text{SU}(2))$

V. *Exceptional Lie Groups*

For each of the five exceptional Lie groups, we have the adjoint representation whose principal orbit type is the maximal tori (T). In addition, we have the following irreducible representations with non-trivial principal isotropy subgroups.

$\mathfrak{g}$	ψ	$\Omega'(\psi)$	(H_ψ^0)
G_2: 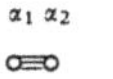	ϕ_1	$\{\pm\theta_1, \pm\theta_2, \pm\theta_3\}$	(SU(3))
F_4: 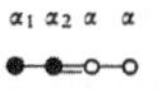	θ_1	$\{\frac{1}{2}(\pm\theta_1 \pm\theta_2 \pm\theta_3 \pm\theta_4), \pm\theta_i\}$	(Spin(8))
E_6: 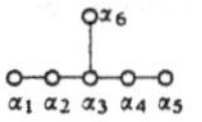	$\phi_1 + \phi_5$		(Spin(8))
E_7: 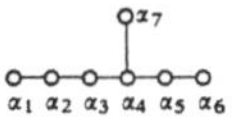	$2\phi_1$		(Spin(8))

Remark. Observe that if $\psi = \psi_1 + \psi_2$, then the respective principal isotropy subgroups of ψ, ψ_1 and ψ_2 are related as follows:

$$(H_\psi) = (H_{\psi_1}) \dot{\cap} (H_{\psi_2}) \quad \text{(intersection in general position).}$$

Hence, it is not difficult to extend the above list for irreducible linear representations to include all linear actions of simple Lie groups.

(B) *G-Admissible and G-Indecomposable Systems of Weights*

Definition. Let G be a compact connected Lie group of rank k and Ω' be a system of non-zero weights defined over a torus group T of rank k. Ω' is called *G-admissible* if there exists an *acyclic* cohomology G-manifold X with $\Omega'(X) = \Omega'$. Furthermore Ω' is called G-decomposable if there exist two *non-trivial* acyclic cohomology G-manifolds X_1 and X_2 such that $\Omega' = \Omega'(X_1) + \Omega'(X_2)$. A G-admissible system of weights Ω' is called G-indecomposable if it is impossible to decompose Ω' into a sum of two non-trivial G-admissible systems of weights.

An obvious *necessary* condition for a system of weights Ω' to be G-admissible is that Ω' is invariant under the action of the Weyl group $W(G)$. However, it is in general far from being sufficient. The following lemma is a slight improvement of such a necessary condition.

Lemma 1. *If Ω' is G-admissible, then Ω' is invariant under the action of $W(G)$ and moreover, to any $w \in \{\Omega' - \Delta(G)\}$ and $\alpha \in \Delta(G)$ with $(w, \alpha) \neq 0$, there exists at least one $w' \in \Omega'$ such*

$$(\alpha - w') \equiv 0 \pmod{w}.$$

Proof. Let X be an acyclic G-manifold with $\Omega'(X) = \Omega'$. (Such a G-space X exists by the assumption that Ω' is G-admissible.) Let w be an arbitrary weights of $\{\Omega' - \Delta(G)\}$. It follows from Lemma 2 of § 2 that there exists an $x \in X$ such that

$$T_x^0 = G_x^0 \cap T = w^\perp \quad \text{is a maximal torus of } G_x^0.$$

Let $\alpha \in \Delta(G)$ be a root with $(w,\alpha) \neq 0$ (non-perpendicular). We claim that $(\alpha|w^{\perp}) \in \{\Delta(G)|w^{\perp} - \Delta(G_x)\}$. Since the case $(\alpha|w^{\perp})=0$ is obvious, we shall prove the above claim for the case $(\alpha|w^{\perp}) \neq 0$ as follows. If the multiplicity of $(\alpha|w^{\perp})$ in $\{\Delta(G)|w^{\perp}\}$ is >1, then it is obvious that $(\alpha|w^{\perp})$ also belongs to $\{\Delta(G)|w^{\perp} - \Delta(G_x)\}$ since every root in $\Delta(G_x)$ has multiplicity one. If the multiplicity of $(\alpha|w^{\perp})$ in $\{\Delta(G)|w^{\perp}\}$ is one, then $(\alpha|w^{\perp}) \in \Delta(G_x)$ if and only if the Lie algebra of G_x, $\mathfrak{g}_x$, containing the eigen-space of α which is obviously impossible because $(w,\alpha) \neq 0$. Hence $(\alpha|w^{\perp}) \notin \Delta(G_x)$, or $(\alpha|w^{\perp}) \in \{\Delta(G)|w^{\perp} - \Delta(G_x)\}$.

Now, it follows from the following equation of Theorem (V.1)

$$\Omega'|w^{\perp} = \Omega'(S_x) + \{\Delta(G)|w^{\perp} - \Delta(G_x)\}$$

that there exists at least one $w' \in \Omega'$ such that

$$w'|w^{\perp} = \alpha|w^{\perp} \in \{\Delta(G)|w^{\perp} - \Delta(G_x)\}$$

or equivalently, $(w' - \alpha) \equiv 0 \pmod{w}$. □

Examples. (1) Let $G = \mathrm{SU}(r+1)$ and $\{\theta_1, \theta_2, \ldots, \theta_{r+1}\}$ be the weight system of the usual representation of $\mathrm{SU}(r+1)$ on $\mathbb{C}^{r+1}$ $(r \geqslant 2)$. Suppose Ω' is a G-admissible system of weights containing $\pm(a\theta_1 + b\theta_2)$, $a > 1$ and $(a,b)=1$. Then, by the above lemma and $(a\theta_1 + b\theta_2, \theta_1 - \theta_3) = a \neq 0$, there exists $w' \in \Omega'$ such that

$$(w' - \alpha) \equiv 0 \pmod{w}$$

or equivalently, there exists a suitable integer k, such that

$$\begin{aligned} w' &= \alpha + k \cdot w = (\theta_1 - \theta_3) + k(a\theta_1 + b\theta_2) \\ &= (k \cdot a + 1) \cdot \theta_1 + (k \cdot b) \cdot \theta_2 - \theta_3 . \end{aligned}$$

Hence, in particular $\{\pm(a\theta_i + b\theta_j)\}$ does not form a G-admissible system of weights.

(2) Let $G = B_r$ (resp. C_r, D_r), $r \geqslant 3$, and $\{\theta_1, \theta_2, \ldots, \theta_r\}$ be the usual orthonormal basis in the Cartan subalgebra of G. Suppose Ω' is a G-admissible system of weights containing $\pm(a\theta_1 + b\theta_2)$, $a > b > 0$ and $(a,b)=1$. Then, again by the above lemma and $(a\theta_1 + b\theta_2, \theta_1 - \theta_3) \neq 0$, there exists k, such that, $w' = (ka+1)\theta_1 + k \cdot b\theta_2 - \theta_3$ also belongs to Ω'.

(C) *Classification of Connected Principal Orbit Types for Topological Actions of Simple Compact Lie Groups on Acyclic Cohomology Manifolds*

We state the main result of § 3 as follows:

Theorem (V.6). *Let G be a simple compact connected Lie group. Let X be a given acyclic cohomology G-manifold with an indecomposable system of non-zero weights $\Omega'(X)$. If the connected principal orbit type, (H_X^0), of X is non-trivial, then there exists a unique irreducible linear G-action ψ with the same non-zero weight system* *and the same connected principal orbit type, i.e., $\Omega'(\psi)=\Omega'(X)$ and $(H^0_\psi)=(H^0_X)$, with the exception of the following undecided possibilities:*

(i) $G=\mathrm{Spin}(11)$, $\Omega'(X)=\{\frac{1}{2}(\pm\theta_1\pm\cdots\pm\theta_5)\}+m\{\pm\theta_i\}\ 2\leqslant m\leqslant 3$,

(ii) $G=\mathrm{Spin}(12)$, $\Omega'(X)=\Omega'(\phi_5)+m\{\pm\theta_i\}\ 2\leqslant m\leqslant 4$.

Lemma 2. *Let G be a simple compact connected Lie group and X be an acyclic cohomology G-manifold. If*

$$\Omega'(X)\supsetneqq\Delta(G)$$

then the connected principal isotropy subgroups is trivial, i.e., $(H^0_X)=\{\mathrm{id}\}$.

Proof of Lemma 2. Let $S\subseteq H^0_X$ be a maximal torus of a principal isotropy subgroup H^0_X. Then it follows from Corollary 1 of Proposition 2 in § 2 that

$$\Omega'(X)|S\equiv\Delta(G)|S-\Delta(H_X)\quad(\mathrm{mod\ zero\ weights}).$$

On the other hand, we assume that $\Omega'(X)\supsetneqq\Delta(G)$, hence

$$\{\Omega'(X)-\Delta(G)\}|S\equiv 0\quad(\mathrm{mod\ zero\ weights}).$$

However, $\{\Omega'(X)-\Delta(G)\}$ is, by assumption, non-empty and invariant under the Weyl group $W(G)$, it is not difficult to see that $\{\Omega'(X)-\Delta(G)\}$ spans the Cartan subalgebra of G. Therefore $\{\Omega'(X)-\Delta(G)\}|S\equiv 0\ (\mathrm{mod}\,0)$ holds only when $S=\{\mathrm{id}\}$, hence H^0_X must be trivial. □

An outline of the proof of Theorem (V.6). The basic idea of the proof of Theorem (V.6) is rather straightforward, it consists of the following steps:

(i) We may assume that $\Omega'(X)$ *does not* contain $\Delta(G)$, for otherwise, it follows from the above Lemma 2 that either $H^0_X=\{\mathrm{id}\}$ or $\Omega'(X)=\Delta(G)$ and $(H^0_X)=(T)$. On the other hand, it follows from Corollary 1 of Proposition 2 of § 2 that

$$\Omega'(X)|H^0_X=\Omega'(\mathrm{Ad}_G|H^0_X)-\Delta(H^0_X).$$

This implies that the following condition is a *necessary condition for the non-triviality of* (H^0_X):

(*) There exists a circle subgroup $S\subseteq T$, such that

$$\Omega'(X|S)\subseteq\Omega'(\mathrm{Ad}_G|S)$$

or, a slightly weaker version which is also easier to check,

(*′) There exists a circle subgroup $S\subseteq T$ such that

$$\dim(\Omega'(X|S))\leqslant\dim(\Omega'(\mathrm{Ad}_G|S)).$$

Since the order of the Weyl group $W(G)$ is usually much larger than the number of roots of G (for example $\mathrm{ord}(W(A_n))=(n+1)!$ as compared to $\#(\Delta(A_n))=n(n+1)$). Hence, it is not difficult to show by Lemma 1 that *almost* all indecomposable weight systems, $\Omega'(X)$, consist of *too many weights* to satisfy condition (*) or even (*'), *except for a few simple possibilities.* Therefore, one needs *only* to examine the remaining few simple possibilities.

(ii) Among those remaining few *possibilities* of indecomposable weight patterns $\Omega'(X)$, at most two or three of them are not realizable by linear actions which deserve special treatment. For such non-linear *possibilities* of indecomposable weight patterns, one may apply the algorithm of Theorem (V.5) to compute their connected principal isotropy subgroups type (H_X^0). If some of them turn out to be trivial, then one may again rule them out.

(iii) After the above two steps of elimination, there are altogether only 15 remaining possibilities of *non-linear* weight patterns that *cannot* be eliminated *solely* by means of weights. For these 15 cases, we proceed to study their orbit structures in detail and then try to examine cohomologically whether it is indeed possible to build an acyclic cohomology manifold with those specific orbit structures. So far, the only *undecided* cases are the possibilities stated in Theorem (V.6).

Proof of Theorem (V.6) for $G=A_n$. In the case $G=A_n$, we usually parametrize the Cartan subalgebra by $(n+1)$ coordinates $(\theta_1,\theta_2,\dots,\theta_{n+1})$ with the relation $\theta_1+\theta_2+\dots+\theta_{n+1}=0$. Then, its Weyl group $W(A_n)$ acts as the full permutation group of the $(n+1)$ coordinates and every weight vector is an integral linear combination of $\{\theta_j\}$. Suppose $\Omega'(X)$ is an indecomposable system of weights. Since $\Omega'(X)$ is invariant under the Weyl group W, we can write $\Omega'(X)$ as the sum of orbits of W as follows

$$\Omega'(X)=W(\pm w_1)+W(\pm w_2)+\cdots.$$

We may assume that $W(\pm w_1)$ is one of the orbits with the largest cardinality and w_1 lies in the Weyl chamber, namely,

$$w_1=a_1\theta_1+\dots+a_k\theta_k, \quad a_1\geqslant\dots\geqslant a_k.$$

Furthermore, since the topological weights are only dependent on their perpendicular hyperplanes, we may assume that

$$(a_1,\dots,a_k)=1 \quad\text{and}\quad a_1\geqslant|a_k|.$$

By Lemma 2, we may assume that $\Omega'(\psi)\cap\Delta(G)=\emptyset$, for otherwise, $\Omega'(\psi)\supseteq\Delta(G)$ and then (H_ψ^0) is either trivial or equal to (T). Suppose $a_1>1$ and $k\neq(n+1)$. Then $(\omega_1,(\theta_1-\theta_{k+1}))\neq 0$ and it follows from Lemma 1 that there exists $w_1'\in\Omega'(\psi)$ such that

$$w_1'=(\theta_1-\theta_{k+1})+l\cdot w_1$$

for a suitable integer l. For most cases, for examples, if $k\leqslant[(n+1)/2]$, the cardinality of $W(w_1')$ is larger than that of $W(w_1)$, which is a contradiction to the choice of w_1. Hence, either $W(w_1)$ already consists of too many weights which makes $\Omega'(X)$ *impossible to satisfy condition* (*'), or $a_1=|a_2|=\dots=|a_k|=1$. Therefore $(H_\psi^0)\neq\{\mathrm{id}\}$ implies that $a_1=|a_2|=\dots=|a_k|=1$.

(i) If $a_1=a_2=\cdots=a_k=1$, then one may assume that $k\leqslant(n+1)/2$ (by using the relation $\theta_1+\theta_2+\cdots+\theta_{n+1}=0$). Notice that the weight systems of the real basic representations are as follows:

$$\Omega'(\varphi_k+\varphi_k^*)=W\{\pm(\theta_1+\cdots+\theta_k)\} \quad \text{if} \quad k<\frac{n+1}{2},$$

$$\Omega'(2\varphi_k)=W\{\pm(\theta_1+\cdots+\theta_k)\} \quad \text{if} \quad k=\frac{n+1}{2}\not\equiv 0 \pmod 2,$$

$$\Omega'(\varphi_k)= \quad W\{(\theta_1+\cdots+\theta_k)\} \quad \text{if} \quad k=\frac{n+1}{2}\equiv 0 \pmod 2.$$

Hence, it follows from Theorem (V.5), Corollary 3 of Theorem (V.5) and I of Table A, that $(H_X^0)\neq\{\mathrm{id}\}$ only when $\Omega'(X)$ is, in fact, the same as one of those listed in Table A—I with possibly the following exceptions:

$$G=A_5 \quad \text{and} \quad \Omega'(X)=W\cdot\{(\theta_1+\theta_2+\theta_3)\}+m\cdot\{\pm\theta_i\}; \qquad (m=0 \text{ or } 1).$$

However, in either case of $m=0$ or 1, a detail computation of their orbit structures will show that the homogeneous space $\frac{\mathrm{SU}(6)}{\mathrm{SU}(3)\times\mathrm{SU}(3)}$ has the same rational cohomology as that of S^{19}, which is a contradiction. Hence $\Omega'=W\{(\theta_1+\theta_2+\theta_3)\}+m\{\pm\theta_i\}$, $m=0,1$ are, in fact, *not admissible* for A_5.

(ii) The remaining possible cases are $w_1=(\theta_1+\cdots+\theta_j-\theta_{j+1}\ldots\theta_k)$. Since $\Omega'(X)\cap\Delta(G)=\emptyset$, we may assume that $k>2$ and $j\geqslant k/2$. Again, it is not difficult to show that $W\cdot(\pm w_1)$ consists of too many weights which makes $\Omega'(X)$ *impossible to satisfy condition* (*'). □

Proof of Theorem (V.6) for $G=B_n, C_n, D_n$ and exceptional Lie groups. The proof of Theorem (V.6) for simple Lie groups other than A_n is essentially the same as that of A_n-case. The first step is to use condition * and Lemma 1 to reduce the *possible candidates* of indecomposable weight system $\Omega'(X)$ with non-trivial connected principal isotropy subgroup type $(H_X^0)\neq\{\mathrm{id}\}$ to a handful of distinguished ones. Among the few remaining candidates of weight patterns, there are the following three kinds:

(i) Those weight patterns that can be realized by linear actions. Then it follows directly from Corollary 1 of Theorem (V.5) that their connected principal orbit types (H_X^0) are the same of those of the corresponding linear actions.

(ii) Those weight patterns which cannot be realized by linear actions, however, the algorithm of Theorem (V.5) applied to them will yield a *trivial* connected principal isotropy subgroups type. Hence, as far as the proof of Theorem (V.6) is concerned, they will not cause any trouble even if some of them turn out to be admissible.

(iii) Finally, there remains the following possible candidates of weight patterns which are non-linear and will yield non-trivial connected principal isotropy subgroup type, i.e., $(H_X^0)\neq\{\mathrm{id}\}$, if some of them happen to be admissible.

(1) $G=\mathrm{SU}(6)$, $\Omega'=\{(\theta_i+\theta_j+\theta_k)\}+m\cdot\{\pm\theta_i\}$, $m=0,1$,
(2) $G=\mathrm{Spin}(11)$, $\Omega'=\{\frac{1}{2}(\pm\theta_1\pm\cdots\pm\theta_5)\}+m\cdot\{\pm\theta_i\}$, $m=0,1,2,3$,
(3) $G=\mathrm{Spin}(13)$, $\Omega'=\{\frac{1}{2}(\pm\theta_1\pm\cdots\pm\theta_6)\}+m\cdot\{\pm\theta_i\}$, $m=0,1$,
(4) $G=\mathrm{Sp}(3)$, $\Omega'=\{(\pm\theta_1\pm\theta_2\pm\theta_3)\}+(m+1)\{\pm\theta_i\}$, $m=0,1$,
(5) $G=\mathrm{Spin}(12)$, $\Omega'=\Omega'(\phi_5)+m\{\pm\theta_i\}$, $0\leqslant m\leqslant 4$.

(iv) In the above five types of non-linear weight patterns with $\underline{m=0}$ it is not difficult to determine the cohomological aspect of the detail "orbit structure" if some of them happen to be admissible. For example, in the case $m=0$, the principal orbit types are respectively the following, which are, in fact, topologically contradictory to the assumption that X is acyclic.

(1) $G=\mathrm{SU}(6)$, $\Omega'=\{(\theta_i+\theta_j+\theta_k)\}\Rightarrow(H_X^0)=(\mathrm{SU}(3)\times\mathrm{SU}(3))$, $F(G)=F(T)$ is acyclic and $\dim\dfrac{\mathrm{SU}(6)}{\mathrm{SU}(3)\times\mathrm{SU}(3)}=19=\dim X-\dim F(G)-1$, which implies $\dfrac{\mathrm{SU}(6)}{\mathrm{SU}(3)\times\mathrm{SU}(3)}$ is a rational cohomology sphere, a contradiction.

(2) $G=\mathrm{Spin}(11)$, $\Omega'=\{\frac{1}{2}(\pm\theta_1\pm\cdots\pm\theta_5)\}\Rightarrow H_X^0=\mathrm{SU}(5)$, $F(G)=F(T)$ is acyclic and $\dim\left(\dfrac{\mathrm{Spin}(11)}{\mathrm{SU}(5)}\right)=31=\dim X-\dim F(G)-1$, which implies that $\dfrac{\mathrm{Spin}(11)}{\mathrm{SU}(5)}$ is a rational cohomology sphere, again a contradiction.

(3) $G=\mathrm{Spin}(13)$, $\Omega'=\{\frac{1}{2}(\pm\theta_1\pm\cdots\pm\theta_6)\}\Rightarrow$ there exists an orbit of the type $\dfrac{\mathrm{spin}(13)}{\mathrm{SU}(6)}$ and the weight system of the SU(6) action on the *slice* has $\Omega'(S_x)=\{(\theta_i+\theta_j+\theta_k)\}$, which is proved in (1) to be impossible. Hence $\{\frac{1}{2}(\pm\theta_1\pm\cdots\pm\theta_6)\}$ is not Spin(13)-admissible.

(4) $G=\mathrm{Sp}(3)$, $\Omega'=\{(\pm\theta_1\pm\theta_2\pm\theta_3)\}+\{\pm\theta_i\}\Rightarrow H_X^0=\mathrm{SU}(3)$, $F(G)=F(T)$ is acyclic, and $\dim\dfrac{\mathrm{Sp}(3)}{\mathrm{SU}(3)}=13=\dim X-\dim F(G)-1$ which again implies that $\dfrac{\mathrm{Sp}(3)}{\mathrm{SU}(3)}$ is a rational cohomology sphere, an obvious contradiction.

(5) Spin(12), $\Omega'=\Omega'(\varphi_5)\Rightarrow H_X^0=\mathrm{SU}(4)$, $F(G)=F(T)$ is acyclic and $\dim\dfrac{\mathrm{Spin}(12)}{\mathrm{SU}(6)}=31=\dim X-\dim F(G)-1$, which again is impossible because $\dfrac{\mathrm{Spin}(12)}{\mathrm{SU}(6)}$ is *not* a rational cohomology sphere.

(v) A more detailed but basically the same method will show that the above five types of non-linear weight patterns with $m=1$ are not admissible either. Hence, the only remaining undecided cases are the following:

$G=\mathrm{Spin}(11)$, $\Omega'=\{\frac{1}{2}(\pm\theta_1\pm\cdots\pm\theta_5)\}+m\cdot\{\pm\theta_i\}$, $m=2$ or 3,

$G=\mathrm{Spin}(12)$, $\Omega'=\Omega'(\varphi_5)+m\cdot\{\pm\theta_i\}$, $2\leqslant m\leqslant 4$. $2\leqslant m\leqslant 4$.

The proof of Theorem (V.6) is thus complete. □

It follows from the above Theorem (V.6) and Corollary 2 of Theorem (V.5), we have the following classification Theorem.

Theorem (V.6′). *Let G be a simple compact connected Lie group and X be an acyclic cohomology G-manifold. If the connected principal orbit type of X, (H_X^0), is non-trivial in the sense $H_X^0 \neq \{\mathrm{id}\}$, then there exists a unique linear G-action φ with the same weight system, i.e., $\Omega(X)=\Omega(\varphi)$, and the same connected principal orbit type, i.e., $(H_X^0)=(H_\varphi^0)$, except possibly the following undecided cases (if they happen to be admissible):*

(i) $G=\mathrm{Spin}(11)$, $\Omega'(X)=\{\tfrac{1}{2}(\pm\theta_1\pm\cdots\pm\theta_5)\}+m\{\pm\theta_i\}$, $2\leqslant m\leqslant 3$,
(ii) $G=\mathrm{Spin}(12)$, $\Omega'(X)=\Omega'(\varphi_5)+m\{\pm\theta_i\}$, $2\leqslant m\leqslant 4$. $2\leqslant m\leqslant 4$.

Conjecture. The above two possibilities are not admissible.

§ 4. Classification of Connected Principal Orbit Types for Actions of (General) Compact Connected Lie Groups on Acyclic Cohomology Manifolds

Let G be a (general) compact connected Lie group and $\mathfrak{g}$ be the Lie algebra of G. It follows from a well-known structural theorem for compact Lie algebras that $\mathfrak{g}$ decomposes uniquely into the direct sum of its center $\mathfrak{g}_0$ and its simple factors $\mathfrak{g}_1, \mathfrak{g}_2, \ldots, \mathfrak{g}_l$, namely

$$\mathfrak{g}=\mathfrak{g}_0\oplus\mathfrak{g}_1\oplus\cdots\oplus\mathfrak{g}_l \qquad (\mathfrak{g}_0 \text{ may be trivial}).$$

Hence, there exists a suitable *finite covering* group $\tilde{G}$ of G such that

$$\tilde{G}=G_0\times G_1\times\cdots\times G_l \qquad (G_0 \text{ may be trivial, i.e., } G_0=\{\mathrm{id}\})$$

where G_0 is a torus group and $G_1,\ldots,G_l$ are simple compact Lie groups with $\mathfrak{g}_1,\ldots,\mathfrak{g}_l$ as their Lie algebra respectively. Hence, in the study of connected principal orbit types, (H_X^0), we may assume without loss of generality that G is itself a product of its connected center G_0 and its simple normal factors $G_1,\ldots,G_l$, i.e.,

$$G=G_0\times G_1\times\cdots\times G_l\,.$$

(A) *Several Reductions*

(i) Suppose that the connected center of G, G_0, is non-trivial, i.e., G is nonsemisimple, and Ψ is an almost effective G-action on an acyclic cohomology manifold X. Let $\Omega_0=\Omega(\Psi|G_0)$ be the weight system of the restriction of Ψ to G_0 and $X_0, X_w, w\in\Omega_0'$ be the following subspaces:

$$X_0=F(G_0,X), \qquad X_w=F(G_0^w,X)$$

where G_0^w is the kernel of $w\in\Omega_0'$. Then X_0, X_w are obviously acyclic cohomology submanifolds of X invariant under G and moreover, the original G-action Ψ is, to a large extent, determined by those restricted G-action on X_0, and X_w, $w \in \Omega'_0$ respectively. For example, if $\dim X_0 = 0$ then

$$\Omega(\Psi) = \sum_{w \in \Omega'_0} \Omega(\Psi | X_w)$$

and it follows directly from Corollary 2 of Theorem (V.5) that

$$(H_X^0) = \dot{\bigcap}\{H_w; w \in \Omega'_0\} \quad \text{(intersection in general position)}$$

where (H_w) is the connected principal orbit type of $\Psi | X_w$.

The general situation, i.e., $\dim X_0 > 0$, can be taken care of by the following generalization of Corollary 2 of Theorem (V.5).

Lemma. *Let G be a given compact connected Lie group and (X, Y), (X_1, Y_1), (X_2, Y_2) be pairs of acyclic cohomology G-manifolds (i.e., Y, Y_1, Y_2 are invariant acyclic submanifolds of X, X_1, X_2 respectively). If their respective systems of non-zero weights satisfy the following equations:*

$$\Omega'(Y) = \Omega'(Y_1) = \Omega'(Y_2)$$

and

$$\Omega'(X) - \Omega'(Y) = \{\Omega'(X_1) - \Omega'(Y_1)\} + \{\Omega'(X_2) - \Omega'(Y_2)\}$$

then their corresponding connected principal orbit types have the following relationship: Let (K) be the connected principal orbit type of Y or Y_1 or Y_2 which are the same by Corollary 1 of Theorem (V.5). Then

$$(H_X^0) = (H_{X_1}^0) \bigcap^{(K)} (H_{X_2}^0)$$

where $\bigcap^{(K)}$ means intersection in general position in K.

Proof. By the above assumption, there are points y, y_1, y_2 in Y, Y_1, and Y_2 respectively such that

$$G_y^0 = G_{y_1}^0 = G_{y_2}^0 = K\,.$$

Let S_y, S_{y_1}, S_{y_2} be the slices at y, y_1, y_2 respectively. Then, it follows from the above equation that $\Omega'(S_y) = \Omega'(S_{y_1}) + \Omega'(S_{y_2})$. Hence, the above Lemma follows from Corollary 2 of Theorem (V.5). ▯

(i) Now, suppose $(H_0), (H_w)$ be the connected principal orbit types of X_0 and X_w respectively. Then

$$(H_X^0) = \bigcap^{(H_0)} \{(H_w); w \in \Omega'_0\}\,.$$

Hence, as far as the principal orbit type is concerned, one may reduce the general case to the case $\dim(G_0) \leqslant 1$.

(ii) Suppose the connected principal isotropy subgroups of X, (H_X^0), are contained in a *normal* subgroup $K \subseteq G$, i.e.,

$$H_X^0 \subseteq K \subseteq G, \quad \text{and } K \text{ is normal.}$$

Then the connected principal isotropy subgroups of the restricted K-space X are the same as that of the G-space X. Hence, the study of principal isotropy subgroups types of topological G-actions is reduced to the study of those cases where the connected principal isotropy subgroups, (H_X^0), are *not contained* in any *proper normal* subgroups of G, or in other words,

$$H_X^0 \subseteq K \subseteq G \quad \text{and} \quad K \text{ normal} \Rightarrow K = G.$$

(iii) Finally, in view of Corollary 2 of Theorem (V.5), one may assume that the system of non-zero weights, $\Omega'(X)$, is *indecomposable*. Furthermore, in case the group G is non-simple, it follows easily from the definition that a splitting weight system is automatically decomposable. Hence, an indecomposable system of weights is necessary non-splittable.

(B) *The Case* $G = G_1 \times G_2$ *is the Product of Two Simple Lie Groups*

Let $G = G_1 \times G_2$ be the product of two simple Lie groups G_1, G_2 and $\mathfrak{h} = \mathfrak{h}_1 \oplus \mathfrak{h}_2$ be a Cartan subalgebra of G and $\mathfrak{h}_1, \mathfrak{h}_2$ be respectively the Cartan subalgebras of G_1, G_2. It is clear that the classification of possible connected principal orbit types for *general* topological G-actions on acyclic manifolds can be reduced to the classification of connected principal orbit types for those G-actions with *indecomposable* system of weights, $\Omega'(X)$, and non-splitting connected principal isotropy subgroups (H_X^0).

Theorem (V.7). *Let* $G = G_1 \times G_2$ *be the product of two simple Lie groups* G_1, G_2 *and* X *be an acyclic cohomology* G*-manifold. If the weight system of* $X, \Omega'(X)$, *is indecomposable and the connected principal isotropy subgroup of* $X, (H_X^0)$, *is not contained in* G_1 *or* G_2, *then either*

(i) $G = \mathrm{SU}(n) \times \mathrm{SU}(m)$, $\Omega'(\Psi) = \Omega'([\mu_n \otimes_{\mathbb{C}} \mu_m]_{\mathbb{R}})$, $(H_\Psi^0) = \mathrm{SU}(n-m) \times T^{(m-1)}$;

or

(ii) $G = \mathrm{Sp}(n) \times \mathrm{Sp}(m)$, $\Omega'(\Psi) = \Omega'(\nu_n \otimes_{\mathbb{H}} \nu_m)$, $(H_\Psi^0) = \mathrm{Sp}(n-m) \times \mathrm{Sp}(1)^m$,

where μ_n, ν_n *are the standard representation of* $\mathrm{SU}(n)$, $\mathrm{Sp}(n)$ *on* $\mathbb{C}^n$ *and* $\mathbb{H}^n$ *(quaternion n-space) respectively.*

As a straightforward consequence of Theorem (V.7), Corollary 2 of Theorem (V.5) and Proposition 3 of § 2, we have the following classification theorem for the possibilities of principal orbit types of topological G-action on acyclic manifolds where G is the product of two simple Lie groups.

Theorem (V.7'). *Let* $G = G_1 \times G_2$ *be the product of two simple Lie groups* G_1, G_2 *and* X *be an acyclic cohomology* G*-manifold. If the connected principal isotropy subgroups type of* $X, (H_X^0)$, *is non-trivial, then there are only the following possibilities:*

(i) *The weight system* $\Omega'(X)$ *is splitting, i.e.,*

$$\Omega'(X) = \Omega'(X)|G_1 + \Omega'(X)|G_2.$$

Hence it follows from Proposition 3 of § 2 that

$$(H_X^0)=(H_X^0\cap G_1)\times(H_X^0\cap G_2)=(H_{X|G_1}^0)\times(H_{X|G_2}^0)$$

which was classified in Theorem (V.6).

(ii) $H_X^0\subseteq G_1$ *or* $H_X^0\subseteq G_2$, *then* $(H_X^0)=(H_{X|G_1}^0)$ *or* $(H_{X|G_2}^0)$, *which was classified in Theorem (V.6).*

(iii) $G=\mathrm{SU}(n)\times\mathrm{SU}(m)$ *(resp.* $\mathrm{Sp}(n)\times\mathrm{Sp}(m)$*) and* $\Omega'(X)=\Omega'([\mu_n\otimes_{\mathbb{H}}\mu_m]_{\mathbb{R}})$ *(resp.* $\Omega'(X)=\Omega'(\nu_n\otimes_{\mathbb{C}}\nu_m)$*)*

$$(H_X^0)=(\mathrm{SU}(n-m)\times T^{m-1})\quad (resp.\ (H_X^0)=(\mathrm{Sp}(n-m)\times\mathrm{Sp}(1)^m)).$$

Proof of Theorem (V.7'). For convenience, we shall assume that $\mathrm{rk}(G_1)\geqslant\mathrm{rk}(G_2)$ and write the weight system $\Omega'(X)$ as the sum of the following three parts

$$\Omega'(X)=\Omega_1+\Omega_2+\tilde{\Omega}$$

where Ω_1,Ω_2 are the subset of those weights which lie in $\mathfrak{h}_1$ and $\mathfrak{h}_2$ respectively and $\tilde{\Omega}$ is the subset of those weights of mixed form. Since $\Omega'(X)$ is assumed to be indecomposable, it must be also non-splittable, namely, $\tilde{\Omega}\neq\emptyset$. Suppose $w_1+w_2\in\tilde{\Omega}$ is such a mixed weight, $w_1\in\mathfrak{h}_1$ and $w_2\in\mathfrak{h}_2$. Then the whole "orbit" of (w_1+w_2) are also weights in $\tilde{\Omega}$, namely

$$W(G)\cdot(w_1+w_2)=\{(\sigma_1(w_1)+\sigma_2(w_2)),\ \sigma_1\in W(G_1),\ \sigma_2\in W(G_2)\}\subseteq\tilde{\Omega}.$$

We claim that except for the two possibilities mentioned in Theorem (V.7), (i.e., $G=\mathrm{SU}(n)\times\mathrm{SU}(m)$, $\Omega'(\tilde{X})=\tilde{\Omega}=\Omega'([\mu_n\otimes_{\mathbb{C}}\mu_m]_{\mathbb{R}})$ and $G=\mathrm{Sp}(n)\times\mathrm{Sp}(m)$, $\Omega'(\tilde{X})=\tilde{\Omega}=\Omega'(\nu_n\otimes_{\mathbb{H}}\nu_m)$), the connected principal isotropy subgroups of X, H_X^0, lie in G_1 and hence contradicts to the assumption that H_X^0 is *not* contained in G_1 or G_2. A detailed proof of the above assertion is rather tedious and it seems to be inevitable to do somewhat case by case checking. However, in principle, it is simply a straightforward application of the algorithm of Theorem (V.5).

Notice that, for a given simple Lie group, there are only a few distinguished orbits (under the action of the Weyl group) whose number of weights is not larger than the number of positive roots. Hence, except for a few particularly simple cases which can easily be checked by the algorithm of Theorem (V.5), $\tilde{\Omega}$ contains at least an orbit $W(G)\cdot(w_1+w_2)$ such that either the number of weights in $W(G_1)\cdot w_1$ is more than that of the positive roots of G_1 or the number of weights in $W(G_2)\cdot w_2$ is more than that of the positive roots of G_2. Let us show, as a typical example, that, in the later case, H_X^0 must be contained in G_1. Suppose $k=\mathrm{rk}(G_2)$. Then it is possible to choose the proceeding $2k$ weights of the algorithm of Theorem (V.5) among those weights in $W(G)\cdot(w_1+w_2)$ as follows: $\alpha_i\in W(G_1)$, $\beta_i,\beta_i'\in W(G_2)$

$$\begin{array}{ll}\gamma_1=\alpha_1(w_1)+\beta_1(w_2), & \gamma_1'=\alpha_1(w_1)+\beta_1'(w_2)\\ \gamma_2=\alpha_2(w_1)+\beta_2(w_2), & \gamma_2'=\alpha_2(w_1)+\beta_2'(w_2)\\ \cdots & \cdots\\ \gamma_k=\alpha_k(w_1)+\beta_k(w_2), & \gamma_k'=\alpha_k(w_1)+\beta_k'(w_2)\end{array}$$

satisfying the following conditions:

$$\{\alpha_1(w_1), \alpha_2(w_1), \ldots, \alpha_k(w_1)\} \quad \text{linearly independent}$$
$$\beta_1(w_2)-\beta_1'(w_2), \beta_2(w_2)-\beta_2'(w_2), \ldots, \beta_k(w_2)-\beta_k'(w_2)$$

also linearly independent.

Then, it is not difficult to see that the maximal torus S of H_X^0 determined by the algorithm is contained in $T_1 \subseteq G_1$, namely,

$$S \subseteq \gamma^\perp \cap \gamma_1'^\perp \cap \cdots \cap \gamma_k^\perp \cap \gamma_k'^\perp \subseteq (\beta_1(w_2)-\beta_1'(w_2))^\perp \cap \cdots \cap (\beta_k(w_2)-\beta_k'(w_2))^\perp = \mathfrak{h}_2^\perp = \mathfrak{h}_1 .$$

We leave the detailed proof of Theorem (V.7′) to the reader.

(C) *The Case that G is a General Semi-Simple Compact Lie Group*

For the case that G is a general semi-simple compact Lie group, we state the following generalization of Theorem (V.7′) without proof. In fact, its proof is only a slight modification of that of Theorem (V.7′).

Theorem (V.7″). *Let G be a semi-simple compact connected Lie group and Ψ be an almost effective topological G-action on an acyclic cohomology manifold. If the weight system of $\Psi, \Omega'(\Psi)$, is indecomposable and the connected principal isotropy subgroups of Ψ, H_Ψ^0, are not contained in any proper normal subgroups of G, then there are only the following two possibilities:*

(i) $G = \mathrm{SU}(n) \times \mathrm{SU}(m)$, $\Omega'(\Psi) = \Omega'([\mu_n \otimes_{\mathbb{C}} \mu_m]_{\mathbb{R}})$, $(H_\Psi^0) = (\mathrm{SU}(n-m) \times T^{m-1})$,

(ii) $G = \mathrm{Sp}(n) \times \mathrm{Sp}(m)$, $\Omega'(\Psi) = \Omega'(\nu_n \otimes_{\mathbb{H}} \nu_m)$, $(H_\Psi^0) = (\mathrm{Sp}(n-m) \times \mathrm{Sp}(1)^m)$.

Remark. With the above neat strong theorem for indecomposable weight system and Corollary 2 of Theorem (V.5), it is not difficult to write down the complete classification of principal orbit types of all possible G-actions on acyclic manifolds for a *given* compact connected semi-simple Lie group G. However, such a statement for all compact connected semi-simple Lie groups *in general* is not very neat and it seems unnecessary to state it as a theorem.

§ 5. A Basic Fixed Point Theorem

In the study of topological (or differentiable) transformation groups on acyclic manifolds, the existence of *fixed-point-free* actions of non-abelian compact connected Lie groups on euclidean spaces represents a major feature where *general topological actions fail to resemble linear actions* in their geometric behavior. However, if the weight systems of the topological actions satisfy some simple condition, then one may modify the proof of [H 5] to obtain the following useful fixed point theorem for *topological* actions on acyclic manifolds which is slightly weaker than that of the differentiable case [cf. Pt. III of H 5].

Theorem (V.8). *Let G be a compact connected Lie group and X be an acyclic $\mathbb{Z}$-cohomology G-manifold. Let $\Omega(X)$ be the weight system of X and*

$$\textstyle\sum_0 = \Delta(G) \setminus \Omega(X), \quad \sum_1 = \{\alpha \in \Delta(G) \text{ and with multi 1 in } \Omega(X)\}.$$

Then there exist a maximal rank subgroup $K \subseteq G$, such that $\Delta(K) \supseteq \sum_0 \cup \sum_1$ and $F(K, X)$ is an acyclic cohomology manifold.

We shall need the following lemma in order to adapt the proof of [H5, III] to the topological setting.

Lemma 1. *Let X be a cohomology $\mathbb{Z}$-acyclic n-manifold with a given topological* SO(3)*-action. If $\Omega'(X)$ consists of only one pair of non-zero weights, then*

$$F(\mathrm{SO}(3), X) = F(\mathbb{Z}_2, F(\mathrm{SO}(2), X))$$

and it is a cohomology $\mathbb{Z}$-acyclic manifold of dimension $(n-3)$.

Proof. Let $Y = F(\mathrm{SO}(2), X)$. Since $\Omega'(X)$ only consists of a single pair of weights, it is clear that Y is a Z-acyclic cohomology manifold of dimension $(n-2)$. We claim that SO(2) acts freely on $(X-Y)$. Since SO(3) contains no normal subgroup (except the identity group), we see that the restricted SO(2)-action must be *effective*. On the other hand, it follows easily from the fact $\dim Y = (n-2)$ that SO(2) acts freely in the neighborhood of the fixed point set Y. Now, suppose on the contrary that there exists a point $x \in (X-Y)$ with $\mathrm{SO}(2)_x = Z_m$, $m > 1$. Let p be a prime factor of m. Then $F(Z_p, X) \supseteq Y \cup \{x\}$ consists of at least two connected components which contradicts a theorem of P.A. Smith that $F(Z_p, X) \sim_{Z_p} pt$. Hence, SO(2) acts freely on $(X-Y)$. As a consequence of the above fact, the SO(3)-action consists of only two types of orbits, namely

$$\mathcal{O}(X) = \{\mathrm{SO}(3), (\mathrm{SO}(2))\}.$$

Now, let $H = Z_2 + Z_2$ be a maximal Z_2-torus of SO(3), say

$$H = \left\{ \begin{pmatrix} \varepsilon_1 & & 0 \\ & \varepsilon_2 & \\ 0 & & \varepsilon_3 \end{pmatrix} ; \ \varepsilon_j = \pm 1, \ \varepsilon_1 \cdot \varepsilon_2 \cdot \varepsilon_3 = 1 \right\}.$$

Then, it is not difficult to see that

$$F(H, X) = F(\mathbb{Z}_2, F(\mathrm{SO}(2), X)) = F(\mathrm{SO}(3), X)$$

and is of dimension $(n-3)$. Since Y is $\mathbb{Z}$-acyclic, and $F(\mathbb{Z}_2, Y)$ is of *codimension one* in Y, it is easy to show that $F(\mathbb{Z}_2, Y)$ is in fact, $\mathbb{Z}$-acyclic rather than $\mathbb{Z}_2$-acyclic. □

The following is a sketch of proof of Theorem (V.8).

The following is a sketch of proof of Theorem (V.8).

Let $T \subseteq G$ be a maximal torus of G and $Y = F(T, X)$. Clearly, the Weyl group $W(G) = N(T)/T$ acts naturally on Y. For a given root $\alpha \in \Delta(G)$, let $T_\alpha = \alpha^\perp$, $G_\alpha = N(T_\alpha)$ and $\tilde{G}_\alpha = G_\alpha / T_\alpha$. Then, it is not difficult to see that the induced $\tilde{G}_\alpha$-action on $Y_\alpha = F(T_\alpha, X)$ has the following weight system, namely,

$$\Omega'(Y_\alpha) = \{\pm\alpha, \text{ with the same multiplicity as in } \Omega'(X)\}.$$

Hence, in case $\alpha \in \sum_1$, the $\tilde{G}_\alpha$-action on Y_α has only one pair of non-zero weights and Lemma 1 applies to show that

$$F(G_\alpha, X) = F(\tilde{G}_\alpha, Y_\alpha) = F(W(G_\alpha), Y) = H_\alpha$$

where $W(G_\alpha) = \mathbb{Z}_2$ and $H_\alpha = F(W(G_\alpha), Y)$ is of codimension one in Y. Therefore, the subgroup W' of $W(G)$ generated by $\{W(G_\alpha); \alpha \in \sum_1\}$ acts as a group generated by *topological reflections* on Y. A modification of the proof of the fixed point theorem for groups generated by *differentiable* reflections in [H 5, III], [H 8] will show that

$$F(W', Y) = \bigcap \{H_\alpha; \alpha \in \textstyle\sum_1\}$$

is also $\mathbb{Z}$-acyclic. Let $K = \bigcap \{G_z; z \in F(W', Y)\}$. Then, it is clear that $K \supseteq T$, $\Delta(K) \supseteq \{\sum_0 \cup \sum_1\}$ and $F(K, X) = F(W', Y)$ is $\mathbb{Z}$-acyclic.

§ 6. Low Dimensional Topological Representations of Compact Connected Lie Groups

In this section, we shall regard a topological G-action, Ψ, on an acyclic cohomology manifold X as a *representation* of G via topological transformation groups, and shall simply call it a *topological representation* of G. Then, it is rather natural to look for the *lowest dimensional topological representation* of a given group G, where $\dim \Psi$ is of course defined to be $\dim X$. For this purpose, let us introduce the following notations:

Definition. For a given compact Lie algebra $\mathfrak{g}$, let

$$L_{\text{top}}(\mathfrak{g}) = \operatorname{Min}\{\dim \Psi\}$$

where Ψ runs through all *effectice topological* transformation groups on cohomology acyclic manifolds with $\mathfrak{g}$ as the Lie algebra. For a given compact Lie group G, let

$$L_{\text{top}}(G) = \operatorname{Min}\{\dim \Psi\}$$

where Ψ runs through all possible *topological effective* G-action on cohomology acyclic manifolds.

Remark. Clearly, one may similarly define $L_{\text{diff}}(\mathfrak{g})$, $L_{\text{diff}}(G)$ and $L_{\text{lin}}(\mathfrak{g})$, $L_{\text{lin}}(G)$ in terms of differentiable actions and linear actions respectively. It is then obvious that

$$L_{\text{top}}(\mathfrak{g}) \leqslant L_{\text{diff}}(\mathfrak{g}) \leqslant L_{\text{lin}}(\mathfrak{g}),$$
$$L_{\text{top}}(G) \leqslant L_{\text{diff}}(G) \leqslant L_{\text{lin}}(G).$$

Hence, the interesting problem here is to estimate $L_{\text{top}}(\mathfrak{g})$ and $L_{\text{top}}(G)$ from below. As an example, we shall generalize the estimate of $L_{\text{diff}}(\text{Spin}(k))$ in [H 14] to the topological case as follows

Theorem (V.9). *Let* $G=\text{Spin}(k)$, *and* $r=[k/2]$. *Then*

$$L_{\text{top}}(G) \geqslant \begin{cases} 2^r & \text{if } k \not\equiv 0 \pmod 4, \\ 2^{r-1}+2r & \text{if } k \equiv 0 \pmod 4. \end{cases}$$

Proof. Let Z_2 be the kernel of the covering homomorphism

$$\text{Spin}(k) \to SO(k)$$

and $T \subseteq \text{Spin}(k)$ be a maximal torus. Let $(\theta_1, \theta_2, \ldots, \theta_r)$ be the usual coordinate for the Cartan subalgebra of $\text{Spin}(k)$, $r=[k/2]$. It is well-known that all weights of $\text{Spin}(k)$ are of the form

$$w = k_1\theta_1 + \cdots + k_r\theta_r \quad \text{with} \quad k_1 \equiv \cdots \equiv k_r \equiv \begin{Bmatrix} 0 \\ \frac{1}{2} \end{Bmatrix} \pmod 1.$$

Let Ψ be an *effective* action of $\text{Spin}(k)$. Then $\Psi|T$ is also *effective*. Let $\Omega'(\Psi)$ be the system of non-zero weights of Ψ. We claim that $\Omega'(\Psi)$ must contain some weights of the form $w=k_1\cdot\theta_1+\cdots+k_r\cdot\theta_r$ with $k_j \equiv \frac{1}{2} \pmod 1$, i.e., half integers. For otherwise one can show that $Y'=F(Z_2, X)$ will be a Q-acyclic manifold of the same dimension as X itself and hence $F(Z_2, X)=X$, Ψ is not effective. Since $\Omega'(\Psi)$ is clearly invariant under the conjugation of the Weyl group, $\Omega'(\Psi)$ must contain the whole "orbit" of such a weight, $W(w)$, which consists of a least 2^r elements if $k=2r+1$ and at least 2^{r-1} elements if $k=2r$. In the case $k \equiv 2 \pmod 4$, $-w$ is not a conjugate weight of w and hence $\Omega'(\Psi)$ must also contain the "orbit" of $W(-w)$. Therefore,

$$\dim(\Psi) \geqslant \text{the number of weights in } \Omega'(\Psi) \geqslant 2^r$$

for the case $k \not\equiv 0 \pmod 4$. In the case $k \equiv 0 \bmod (4)$, the center of $\text{Spin}(k)$ is Z_2+Z_2 which has three Z_2 subgroups. Suppose $\Omega'(\Psi)$ only consists of a single orbit $W(w)$ of the above type with exactly 2^{r-1} elements. Then the same reason will show that $\text{Ker}(\Psi)$ is another Z_2 subgroup which contradicts to the assumption that Ψ is effective. Hence, $\Omega'(\Psi)$ either contains one more orbit or the orbit $W(w)$ consists of more than 2^{r-1} elements. It is then easy to see that

$$\dim(\psi) \geqslant \text{number of elements in } \Omega'(\Psi) \geqslant 2^{r-1}+2r$$

for the case $k \equiv 0 \pmod 4$. □

Remark. It is not difficult to see that

$$L_{\text{lin}}(\text{Spin}(k)) = \begin{cases} 2^r, & k \equiv \pm 1, \pm 2, 4 \pmod 8, \\ 2^{r+1}, & k \equiv \pm 3 \pmod 8, \\ 2^{r-1} + 2r, & k \equiv 0 \pmod 8). \end{cases}$$

Hence, the above estimates are best possible for the case $k \equiv 0, \pm 1, \pm 2 \pmod 8$. Namely

$$L_{\text{top}}(\text{Spin}(k)) = L_{\text{lin}}(\text{Spin}(k))$$

for $k \equiv 0, \pm 1, \pm 2 \pmod 8$.

As a preliminary step to the complete determination of $L_{\text{top}}(\mathfrak{g})$, we shall first compute the case of simple Lie algebras. For convenience, we shall introduce the following notations:

$\dim \Omega(\Psi) = \dim(\Psi) =$ the number of weights counted with multiplicity.

$\dim \Omega'(\Psi) =$ the number of non-zero weights counted with multiplicity.

Theorem (V.10). *Let $\mathfrak{g}$ be a simple compact Lie algebra and Ψ be a topological G-action on an acyclic cohomology n-manifold X with $\mathfrak{g}$ as the Lie algebra of G. If*

$$\dim \Omega'(\Psi) \leqslant L_{\text{lin}}(\mathfrak{g}),$$

then the fixed point set $F(G,X)$ is also an acyclic cohomology manifold. Moreover, with the only exceptional case of $\mathfrak{g} = A_2$, $\Omega'(\Psi) = \Delta(A_2)$, $\operatorname{codim} F(G,X) = 8$, one has

$$\operatorname{codim} F(G,X) = L_{\text{lin}}(\mathfrak{g}),$$
$$\Omega'(\Psi) = \Omega'(\varphi)$$

where φ is the unique lowest dimensional real linear representation of $\mathfrak{g}$ given by the following table:

$\mathfrak{g}$	$\dim \varphi$	$\Omega'(\varphi)$
$A_n, n \neq 1, 3$	$2(n+1)$	$\{\pm\theta_i, i = 1, \dots, (n+1)\}$
$B_n, n \geqslant 1$	$2n+1$	$\{\pm\theta_i; i = 1, \dots, n\}$
$C_n, n \geqslant 3$	$4n$	$2 \cdot \{\pm\theta_i; i = 1, \dots, n\}$
$D_n, n \geqslant 3$	$2n$	$\{\pm\theta_i; i = 1, \dots, n\}$
E_6	54	$\{\pm(\lambda + \theta_i), \pm(\theta_i + \theta_j); i < j\}$
E_7	112	$2 \cdot \{\pm(\theta_i + \theta_j); i < j\}$
E_8	246	$\Delta(E_8)$
G_2	7	$\{\pm\theta_i, i = 1, 2, 3\}$
F_4	26	$\{\pm\theta_i, \frac{1}{2}(\pm\theta_1 \pm\theta_2 \pm\theta_3 \pm\theta_4)\}$

Corollary. *If* $\mathfrak{g}$ *is a simple compact Lie algebra, then it follows easily from Theorem (V.10) that*

$$L_{\text{top}}(\mathfrak{g}) = L_{\text{diff}}(\mathfrak{g}) = L_{\text{lin}}(\mathfrak{g}).$$

Proof of Theorem (V. 10). We divide the proof into the following five cases:

(i) *The case* $\mathfrak{g}=A_n\,(n\geqslant 3)$, $D_n\,(n\geqslant 3)$ *or* E_6*:* In this case, the lowest dimensional representation φ has no zero weights and $\Omega'(\varphi)$ consists of a single orbit of $W(G)$ which is the unique orbit with the smallest number of elements: Hence, it is clear that $\Omega'(\Psi)=\Omega'(\varphi)$, and $F(G,X)=F(T,X)$ is also an acyclic cohomology submanifold with

$$\operatorname{codim} F(G,X) = \dim\varphi = L_{\text{lin}}(\mathfrak{g}).$$

(ii) $\mathfrak{g}=B_n$ $(n\geqslant 1)$; E_8*:* In this case, $\Omega'(\varphi)$ again consists of a single orbit of $W(G)$ which is the unique orbit with the smallest number of elements. However, $\Omega'(\varphi)\subseteq\Delta(\mathfrak{g})$ and $\Omega(\varphi)$ does have zero weights with multiplicity 1 and 8 respectively. Hence, it is easy to show that $\Omega'(\Psi)=\Omega'(\varphi)$ and then it follows from Theorem (V.8) that $F(G,X)=F(W(G),F(T,X))$ is also an acyclic cohomology manifold with $\operatorname{codim} F(G,X)=\dim\varphi=L_{\text{lin}}(\mathfrak{g})$.

(iii) $\mathfrak{g}=C_n$ $(n\geqslant 3)$, E_7*:* In this case, $\Omega'(\varphi)$ consists of the smallest orbit of $W(G)$ with multiplicity 2. Since all the other orbits of $W(G)$ consists of too many weights, it is clear that

$$\Omega'(\Psi) = \left\{\begin{matrix} l\cdot\{\pm\theta_i\} & \text{if} \quad \mathfrak{g}=C_n,\, n\geqslant 3 \\ l\cdot\{\pm\theta_i+\theta_j\} & \text{if} \quad \mathfrak{g}=E_7 \end{matrix}\right\}, \qquad l=1,2.$$

We shall show that l must be equal to 2. Since the proofs for C_n and E_7 are essentially the same, we shall only prove the C_n case. It follows from the fact that

$$\Delta(G_y) \supseteq \Delta(G)\setminus\Omega'(\Psi) = \{\pm\theta_i \pm\theta_j\}$$

for any point $y\in F(T,X)$; G_y must be in fact equal to G itself. Let $x\in X$ be a point fixed under the corank one subtorus $\theta_1^\perp$. Then, it follows easily from Proposition 2, § 2, that $G_x=\mathrm{Sp}(n-1)$ and l must be equal to 2.

(iv) $\mathfrak{g}=G_2$, F_4*:* In this case $\Omega'(\varphi)$ consists of the short roots of $\mathfrak{g}$, and $\Delta(\mathfrak{g})$ splits into two orbits of equal size, i.e., the long roots and the short roots. Hence, it follows easily from the condition $\dim\Omega'(\Psi)\leqslant L_{\text{lin}}(\mathfrak{g})$ that either

$$\Omega'(\Psi) = \{\text{long roots}\} \quad \text{or} \quad \Omega'(\Psi) = \{\text{short roots}\} = \Omega'(\varphi).$$

Similar reason as that of (iii) will show that the case $\Omega'(\Psi)=\{\text{long roots}\}$ is impossible. Then, again, it follows from Theorem (V.8) that $F(G,X)=F(W(G),F(T,X))$ is also an acyclic manifold and $\operatorname{codim} F(G,X)=\dim\varphi=L_{\text{lin}}(\mathfrak{g})$.

(v) $\mathfrak{g}=A_2$. In this case $L_{\text{lin}}(\mathfrak{g})=6$ and there are only two orbits of $W(G)$ with only three pairs of weights, namely, $\Delta(\mathfrak{g})$ or $\{\pm\theta_i\}$: As in the proceeding cases, it is clear that $F(G,X)$ is also acyclic and $\operatorname{codim} F=8$, or 6 respectively. □

With a strong theorem such as Theorem (V.10) for the case of simple Lie groups, it is not difficult to proceed to prove the following theorem for the case of general compact connected Lie groups. Namely,

Theorem (V.10′). *For any compact Lie algebra* $\mathfrak{g}$, $L_{\mathrm{top}}(\mathfrak{g}) = L_{\mathrm{lin}}(\mathfrak{g})$.

We refer the reader to p. 362—366 of [H 16] for the proof of above theorem.

§ 7. Concluding Remarks Related to Geometric Weight System

(A) *Local Theory of Topological Transformation Groups*

Parallel to most results of this chapter on *global* geometric behavior of actions on acyclic cohomology manifolds, there are corresponding results in the *local theory* of topological transformation groups. Generally speaking, corresponding to each theorem asserting a certain definite relationship between the (global) geometric behavior of the acyclic cohomology G-manifold X and its weight system $\Omega(X)$, there is a local theorem asserting the same type of relationship between the local geometric behavior around the orbit $G(x)$ and the local weight system of the G_x-action on the slice S_x, $\Omega(S_x)$. Technically, such local theorems can be obtained from the corresponding global theorems (for actions on acyclic cohomology manifolds) by a systematic localization of everything involved, applied to the G_x-action on the slice S_x at x. For example, let us state the localized version of Theorem (V.2), Theorem (V.5), Theorem (V.6), and Theorem (V.9) as follows.

Theorem (V.$\bar{2}$). *Let X be a cohomology G-manifold and $x \in X$ be a point with $G_x^0 = \mathrm{SO}(n)$ (resp. $\mathrm{SU}(n)$, $\mathrm{Sp}(n)$). If the local weight system at x, $\Omega(S_x)$ is*

$$\Omega'(S_x) = \{\pm\theta_i, \text{ with multi } l\}.$$

Then there exists an invariant neighborhood N of the orbit $G(x)$ such that

(i) *all connected isotropy subgroups of points in N are conjugate to the standard $\mathrm{SO}(k)$ (resp. $\mathrm{SU}(k)$, $\mathrm{Sp}(k)$) for suitable $k \leqslant n$,*

(ii) *the connected principal isotropy subgroup type (H) is non-trivial when and only when $l \leqslant (n-2)$ (resp. $l < (n-2)$, $l \leqslant 2(n-1)$) and $H = \mathrm{SO}(n-l)$ (resp. $\mathrm{SU}(n-l)$, $\mathrm{Sp}(n-l/2)$, l must be even when $G = \mathrm{Sp}(n)$ and $l < 2n$).*

Theorem (V.$\bar{5}$). *Let X be a connected, cohomology G-manifold, and (H_X^0) be the connected principal orbit type of X. Let $T \subseteq G_x^0$ be a maximal torus of the connected isotropy subgroup of a point $x \in X$, and $\Omega'(S_x)$ be the local system of non-zero weights at x. Up to a conjugation, one may assume that $S = (H_X \cap T)^0$ is a maximal torus of H_X^0. Then there exist a sequence of non-zero weights $w_1, \ldots, w_k \in \Omega'(S_x)$*

together with a sequence of decreasing subtori $T=S_0 \supsetneq S_1 \supsetneq \cdots \supsetneq S_k = S$ *satisfying the following recursive conditions:*

$$S_0 = T,\ w_1 \in \{\Omega'(S_x) - \Omega'(\mathrm{Ad}_{G_x})\} \neq \emptyset$$
$$S_1 = w_1^{\perp},\ w_2 \in \{\Omega'(S_x)|S_1 - \Omega'(\mathrm{Ad}_{G_x}|S_1)\} \neq \emptyset$$
- -
$$S_i = S_{i-1} \cap w_i^{\perp},\ w_{i+1} \in \{\Omega'(S_x)|S_i - \Omega'(\mathrm{Ad}_{G_x}|S_i)\} \neq \emptyset$$
- -
$$S_k = S_{k-1} \cap w_k^{\perp},\ \{\Omega(S_x)|S_k - \Omega'(\mathrm{Ad}_{G_x}|S_k)\} = \emptyset \quad (empty).$$

Conversely, suppose there exist a sequence of non-zero weights and a sequence of decreasing subtori from T to S satisfying the above recursive conditions. Then S is a maximal torus of H_X^0. *Moreover the root system of* H_X^0, $\Delta(H_X^0)$ *is given by the following equation:*

$$\Delta(H_X^0) = \{\Omega'(\mathrm{Ad}_{G_x}|S_k) - \Omega'(S_x)|S_k\}.$$

Theorem (V.$\bar{6}$). *Let X be a connected cohomology G-manifold and* $x \in X$ *be a point with* G_x^0 *a simple compact Lie group. If the local system of non-zero weights at x,* $\Omega'(S_x)$ *is indecomposable and the connected principal orbit type of X,* (H_X^0), *is non-trivial, then there exists a unique irreducible linear representation* ψ *of* G_x^0 *with the same non-zero weight system and the same connected principal orbit type, i.e.,* $\Omega'(S_x) = \Omega'(\psi)$ *and* $(H_X^0) = (H_\psi^0)$, *except the following undecided possibilities:*

(i) $G_x^0 = \mathrm{Spin}(11)$, $\Omega'(S_x) = \{\tfrac{1}{2}(\pm\theta_1 \pm \cdots \pm \theta_5)\} + m\{\pm\theta_i\}$, $2 \leqslant m \leqslant 3$,
(ii) $G_x^0 = \mathrm{Spin}(12)$, $\Omega'(S_x) = \Omega'(\phi_5) + m\{\pm\theta_i\}$, $2 \leqslant m \leqslant 4$.

Theorem (V.$\bar{9}$). *Let X be a cohomology manifold with a given effective topological G-action. If there exists a point* $x \in X$ *with* $G_x \cong \mathrm{Spin}(k)$. *Then*

$$\dim X \geqslant \begin{cases} 2^r & \text{if } k \not\equiv 0 \,(\mathrm{mod}\, 4) \\ 2^{(r-1)} + 2r & \text{if } k \equiv 0 \,(\mathrm{mod}\, 4) \end{cases} \quad r = [k/2] = \mathit{rank}\ G_x^0.$$

Remark. The local linearity of the differentiable slice theorem readily introduces the theories of vector bundles, characteristic classes and linear representations as powerful tools in the study of orbit structure of differentiable actions. We refer the reader to [H 10, H 5 I, II, H 18, G 6] for such applications. Now, with *local weight system* as a functional substitute for the "local linearity", it is then possible to make the topological slice theorem more effectively useful than before.

(B) *Splitting Principle and Generalizations of Geometric Weight System*

In the characteristic class theory, one uses the linear splitting of complex representations of *torus groups* to get the splitting of characteristic classes. In the case of topological transformations of torus groups, the splitting at the geometric level is out of question. However, the results of Chapter IV demonstrate that various kinds of splittings at the level of generalized characteristic classes still hold. (Such splittings are actually reformulated in terms of the linearity of structural ideals of equivariant cohomology algebras.) With the help of such splitting theorems on equivariant cohomology algebras, it is sometimes possible to organize and condense the *cohomological structural data* of *topological torus group actions on spaces of a given type* into some kind of neat genetic code. For example, the geometric weight system, $\Omega(X)$, is exactly such a genetic code for acyclic cohomology manifold, X, with a given action of torus group G. And Theorem (V.1) is exactly the central structural theorem that enables us to organize and condense the cohomological structural data of acyclic cohomology G-manifolds X (G is a torus) into its genetic code—the weight system $\Omega(X)$. In later chapters, we shall prove similar structural theorems for topological actions of torus groups on manifolds of various cohomology types which will then enable us to define geometric weight systems for such actions. For example, let us mention here the following straightforward generalization of Theorem (V.1):

Theorem (V.1″). *Let G be a torus group and X be a cohomology manifold with a given G-action. Suppose that the fixed point set $F(G,X)$ is non-empty and $\pi_q(X)\otimes\mathbb{Q}=0$ for all even $q>0$. Then, it follows from Theorem* (IV.5), *that the $F(K,X)$'s are connected cohomology submanifolds for all subtori $K\subseteq G$. Hence, in particular, $F(G,X)$ is connected and the local weight system $\Omega_x(X)$, $x\in F(G,X)$, is independent of the choice of x which we shall simply define it to be the weight system of X, and denoted by $\Omega(X)$. Moreover, the orbit structure of X can be read off from the weight system $\Omega(X)$ as follows:*

(i) *a subgroup $K\in\mathcal{O}^0(X)$ if and only if K is of the form*

$$K = w_{i_1}^{\perp}\cap\cdots\cap w_{i_k}^{\perp} \qquad (K=G \text{ if } k=0)$$

for suitable subcollection of weights in $\Omega(X)$,

(ii) *there is a bijection between the family of F^0-varieties, $\{Y\}$, and the set of connected isotropy subgroups, $\mathcal{O}^0(X)$, given by $Y\leftrightarrow G_Y^0$, and*

$$\dim Y = \textstyle\sum_{w^{\perp}\supseteq G_Y^0} \operatorname{multi}(w).$$

Notice that, for most results of this chapter, one need only the *dimension* and the *connectedness* of $F(K,X)$ rather than the *acyclicity* of $F(K,X)$. Hence, essentially the same proofs will show that those results such as Propositions 1, 2, and 3 of § 2, Theorems (V.2), (V.5), (V.6), (V.6′), (V.7), (V.7′), and (V.9) are directly generalizable to the case of cohomology G-manifold X under the condition

$$\pi_q(X)\times\mathbb{Q} = 0 \quad \text{for all even} \quad q>0$$

and $F(T,X)\neq\emptyset$ (non-empty) for a maximal torus T of G.

Intuitively speaking, the *geometric weight system* of a given G-space X is simply a neat book-keeping device for the *cohomological aspects* of the orbit structure of the *restriction* to a maximal torus $T\subseteq G$. Usually, one need a rather strong structural theorem for actions of *torus groups* to set up such a neat book-keeping device. Once the orbit structure of the restricted T-action on the given G-space X is *proved* to be sufficiently simple so that it can be well-organized by a neat book-keeping device to be called the geometric weight system of X, then the central problem relating to geometric weight system is to *recover* as much as possible of the *orbit structure of the G-action on X* from that of the *restricted T-action on X*.

(C) *Transformation Groups on Acyclic Cohomology Manifolds with Simple Orbit Structures*

Generally speaking, the orbit structure of topological G-action on acyclic manifolds are often very complicated, e.g., the equivariant embedding theorem of Mostow-Palais shows that all imaginable topological G-spaces are actually sub-G-spaces of some linear G-spaces, which fully demonstrates that the orbit structure of *linear* G-spaces are already as complicated as one can imagine. Hence, it is sensible to start our investigation of orbit structures of topological G-actions on acyclic manifolds with a preliminary study of those special G-actions on acyclic manifolds whose orbit structures are simple. For example, it is rather natural to consider the following three kinds of simplifications in orbit structures as testing cases, namely,

(i) the number of orbit types is not large as compared to the rank of G,

(ii) the dimension of the G-manifold X is not large as compared to the dimension of the group G,

(iii) the dimension of the orbit space, X/G (or equivalently, the codimension of the principal orbits), is small.

A natural approach to this type of problem will be the following steps:

(1) Explicit determination of those *linear actions* satisfying the *given geometric condition.* They will serve as the *linear models* for further investigation.

(2) Analyze the relationship between the *given geometric condition* and the *special pattern* of the weight system of *linear models.*

(3) Try to show that the *weight system* of those *topological actions* on acyclic manifolds satisfying the given geometric condition must be also of similar (or identical) *special pattern* as that of the linear models.

(4) Use the *weight system* as the "substitute" of linear models, try to deduce other geometric characteristics that are *implied* by the *given geometric condition.*

We refer the readers to [H16] for a more thorough discussion of topological transformation groups with simple orbit structures. However, it is interesting to note that most transformation groups on acyclic manifolds with simple orbit structures will automatically have *non-trivial connected principal isotropy subgroups*, and hence the classification results of § 3 and § 4 will be a great help in the study of such group actions.

Chapter VI. The Splitting Theorems and the Geometric Weight System of Topological Transformation Groups on Cohomology Projective Spaces

In this chapter, we take those manifolds of the cohomology type of projective spaces as the testing spaces for the study of topological transformation groups. From the cohomological point of view, the projective spaces certainly have the simplest, and yet non-trivial, cohomology algebras, namely, truncate polynomial rings. Geometrically, the so-called projective transformation groups which are induced by the linear transformation groups still provide abundant interesting examples that we shall again call them "*linear models*". In other words, projective spaces, endowed with a simple cohomology structure and an abundance of transformation groups, provide the ideal setting for the study of the cohomology theory of transformation groups.

In the study of transformation groups on cohomology projective spaces, those elementary abelian groups (i.e., torus and $\mathbb{Z}_p$-torus) again play the crucial rôle. The central results of this chapter are those structure theorems which state that the *cohomological aspects of orbit structures of elementary abelian transformation groups on cohomology projective spaces are the same as those of suitable linear models.* Conceptually, it is interesting to note that such cohomological structure theorems are formulated and proved in the setting of splitting theorems of the structure of equivalent cohomology. In the case of linear models, the splitting at the level of equivariant cohomological structure is a *consequence of the splitting at the geometric level*, which is itself a well-known consequence of the Schur lemma. Hence, the cohomological structure theorems of this chapter may be regarded as the generalizations of the Schur lemma for topological transformation groups on cohomology projective spaces. In the theory of linear transformation groups, the Schur lemma is exactly the basis for setting up the well-known invariant called *weight system.* Correspondingly, the structure theorems also provide the basis for introducing a similar invariant that we shall call the *geometric weight* system.

In § 1, we consider the case of torus group actions on cohomology complex projective spaces. The results are yet unpublished but have long served as the prototype for further developments in the cohomology theory of transformation groups such as [H15, H17, H20]. Technically, the structure theorem for this case, Theorem (VI.1), is but a straightforward specialization of the general splitting theorems of Chapter IV. Its novelty lies only in its formulation which, in fact, was what first led to the discovery of those general splitting theorems.

In § 2, we investigate the case of torus group actions on cohomology quaternonic projective spaces. The results are taken mainly from a joint paper with J. C. Su [H 20]. It is rather amusing to note that such a minute difference in the degree of the generator (i.e., 4 instead of 2) not only makes the corresponding structure theorem, Theorem (VI.6), considerably harder to prove but is also responsible for some interesting new features and examples (cf. § 2).

The case of $\mathbb{Z}_p$-tori (p odd primes) acting on Z_p-cohomology complex (resp. quaternionic) projective spaces is briefly discussed in § 3. The results are parallel to the characteristic zero case of § 1 and § 2 and the proofs need only minor modifications. However, the case of $\mathbb{Z}_2$-tori acting on $\mathbb{Z}_2$-cohomology real, complex, or quaternionic projective spaces are quite different from the other cases. They not only have interesting new features (cf. Theorems (VI.7), (VI.8), and (VI.9)), but also require new proofs involving the Steenrod square operations. We discuss the results of [H 19] concerning $\mathbb{Z}_2$-tori actions on cohomology projective spaces in § 4.

§ 1. Transformation Groups on Cohomology Complex Projective Spaces

In this section, we take those cohomology manifold X of *the rational cohomology type of complex projective spaces* as the testing spaces for the study of transformation groups. We shall call them cohomology complex projective spaces (abbreviated as CCP-spaces) and we shall use the rational field k as coefficients of all cohomology algebras in this section except specified to be otherwise. It was first proved by J.C. Su [S 14] that the connected components of the fixed point set of a circle group action on a CCP-space are again CCP-spaces. One needs the following structure theorem for torus group actions on CCP-spaces to set up a similar geometric weight system for CCP-spaces as that for the case of acyclic manifolds in Chapter V.

(A) *A Structural Splitting Theorem for Tori Actions on CCP-Spaces*

Theorem (VI.1). *Let G be a torus group and X be a CCP^n with a given G-action. Let ξ be an arbitrary lifting of the generator ξ_0 of $H^*(X)$ into $H^*_G(X)$. Then $H^*_G(X) \cong R[\xi]/\langle f(\xi)\rangle$ where*

$$f(\xi) = (\xi - w_1)^{k_1} \dots (\xi - w_s)^{k_s}, \quad w_j \in H^2(B_G)$$

splits into the product of linear factors. Correspondingly, the fixed point set $F(G,X)$ consists of s connected components $\{F^j, 1 \leqslant j \leqslant s\}$ such that

(i) *$F^j \sim CP^{(k_j-1)}$ and $\iota_j^*: H^*_G(X) \to H^*_G(q_j)$, $q_j \in F^j$, maps ξ to $w_j \in R$,*

(ii) *the system of local weights at F^j, Ω_j, is given by $\Omega_j = \{\pm(w_j - w_i), k_i (i \neq j); 0, 2(k_j - 1)\}$,*

(iii) *for a given point $x \in X$ with $F(x) \cap F(G,X) = F^{j1} + \dots + F^{jl}$, the connected isotropy subgroup G_x^0 is given by $w_{j1} = \dots = w_{jl}$ and $F(x) \sim CP^m$, $m = (k_{j1} + \dots + k_{jl} - 1)$.*

Proof. Since the cohomology of $X, H^*(X) \cong k[\xi_0]/\langle \xi_0^{n+1} \rangle$, $\deg \xi_0 = 2$, consists of *only* even degree elements, it is easy to see that the Serre spectral sequence of the fibration $X \to X_G \to B_G$ has no non-zero differentials. Hence $E_2 = E_\infty = H^*(B_G) \otimes H^*(X)$ and $H_G^*(X) \to H^*(X)$ is surjective. Let ξ be an arbitrary lifting of ξ_0 into $H_G^*(X)$. Then $1, \xi, \xi^2, \dots, \xi^n$ form an R-module basis of $H_G^*(X)$ and $H_G^*(X) \cong R[\xi]/\langle f(\xi) \rangle$, where

$$f(\xi) = \xi^{n+1} + c_1 \xi^n + c_2 \xi^{n-1} + \cdots + c_{n+1} = 0$$

is the defining structural equation of $H_G^*(X)$. It follows from a direct application of Theorem (IV.1) that all roots of the above equation $f(\xi) = 0$ are "rational", i.e., $f(\xi) = (\xi - w_1)^{k_1} \cdot (\xi - w_2)^{k_2} \dots (\xi - w_s)^{k_s}$, $w_j \in H^2(B_G)$, and correspondingly, the fixed point set $F(G, X)$ consists of s connected components $\{F^j, 1 \leqslant j \leqslant s\}$ such that $\xi | H_G^*(q_j) = w_j$ for $q_j \in F^j$.

Let ξ_j be the generator of $H^*(F^j) \cong k[\xi_j]/\langle \xi_j^{k_j} \rangle$, and ι^* be the restriction homomorphism: $H_G^*(X) \to H_G^*(F) \cong \sum_{j=1}^s H_G^*(F^j) = \sum_{j=1}^s R \otimes H^*(F^j)$. Then it is not difficult to see that ι^* is injective and

$$\iota^*(\xi) = (\xi_1 + w_1, \xi_2 + w_2, \dots, \xi_s + w_s).$$

Let $f_j = (0, \dots 0, \xi_j^{(k_j - 1)}, 0, \dots, 0)$ be the fundamental cohomology class of F^j, and $I(f_j) = I_{f_j}(X, F)$ be the ideal of those $a \in R$ with $a \otimes f_j \in \operatorname{Im}(\iota^*)$ [cf. § 2—(c) of Ch. IV]. Suppose a is an arbitrary element of $I(f_j)$ and $\iota^*(g(\xi)) = a \otimes f_j$. Then $\iota^*(g(\xi) \cdot (\xi - w_j)) = (a \otimes f_j) \cdot (1 \otimes \xi_j) = 0$ and it follows from the injectivity of ι^* that $g(\xi) \cdot (\xi - w_j)$ is divisible by $f(\xi)$, so that $g(\xi)$ is divisible by $\dfrac{f(\xi)}{(\xi - w_j)}$. On the other hand, it is easy to see that $\iota^* \dfrac{f(\xi)}{(\xi - w_j)} = a_j \otimes f_j$ where $a_j = \prod_{i \neq j} (w_j - w_i)^{k_i}$ and

$$\iota^*\left(h(\xi) \cdot \frac{f(\xi)}{(\xi - w_j)}\right) = (h(\xi) | H_G^*(q_j)) \cdot a_j \otimes f_j.$$

Hence $I(f_j)$ is generated by $a_j = \prod_{i \neq j} (w_j - w_i)^{k_i}$ and it follows directly from Theorem (IV.10) that the system of local weights at F^j, Ω_j, is given by

$$\Omega_j = \{\pm(w_j - w_i), k_i (i \neq j); 0, 2(k_j - 1)\}.$$

Let $Y = F(x)$ be the F^0-variety of x and $Y \cap F(G, X) = F^{j_1} + \cdots + F^{j_l}$, and $\bar{\xi}$ be the restriction of ξ to $H_G^*(Y)$. Then, it is clear that $Y \sim CP^m$, $m = (k_{j1} + \cdots + k_{jl} - 1)$ and $H_G^*(Y) \cong R[\xi]/\langle \bar{f}(\bar{\xi}) \rangle$, where

$$\bar{f}(\bar{\xi}) = (\bar{\xi} - w_{j1})^{k_{j1}} \dots (\bar{\xi} - w_{jl})^{k_{jl}}.$$

On the other hand, it is clear that the generic isotropy subgroup of X, $G_X^0 = \operatorname{Ker}^0(X)$, is given by putting in Ω_1 equal to zero, i.e., $(w_1 - w_2) = (w_1 - w_3) = \cdots = (w_1 - w_s) = 0$, or equivalently,

$$w_1 = w_2 = \cdots = w_s.$$

Hence, if we apply the above result to $Y=F(x)$ instead of X, we get that $G_x^0=G_Y^0$ is given by $w_{j1}=\cdots=w_{jl}$, and the proof is thus complete. □

Remarks. (i) In case X is a CCP^n over $\mathbb{Z}$ (instead of over k), then one may use integral coefficients and $H_G^*(X)\cong H^*(B_G,\mathbb{Z})[\xi]/\langle f(\xi)\rangle$ with $f(\xi)=(\xi^{n+1}+c_1\xi^n+\cdots+c_{n+1})\in H^*(B_G,\mathbb{Z})[\xi]$. The above theorem asserts that $f(\xi)$ splits in $H^*(B_G,k)[\xi]$. Hence, by the famous Gauss lemma, $f(\xi)$ also splits in $H^*(B_G,\mathbb{Z})[\xi]$, i.e.,

$$f(\xi)=(\xi-w_1)^{k_1}\cdots(\xi-w_s)^{k_s} \quad \text{with} \quad w_j\in H^2(B_G,\mathbb{Z}).$$

(ii) Notice that the systems of local weights at F^j, Ω_j, in the above theorem are by definition (cf. § 1—c of Ch. V) rational weights which only have definite directions but no definite lengths. However, in the special case that X is a differentiable G-manifold with $H^*(X;\mathbb{Z})\cong\mathbb{Z}[\xi_0]/\langle\xi_0^{n+1}\rangle$, then $w_j\in H^2(B_G,\mathbb{Z})$ are integral weight vectors and the integral weight systems of the local (linear) representation of G at F^j are also defined. Hence, it is interesting to ask whether the integral weight system of the local representation of G at F^j is still given by the same formula in terms of the integral weights $w_j\in H^2(B_G,\mathbb{Z})$. It can be proved to be true for many cases but there also exists some counter-examples [P 2] in the case G is a circle group. It seems plausible that it holds in general for the case $\mathrm{rk}(G)\geqslant 2$ or 3.

(B) *Linear Models and the Geometric Weight System for Actions on CCP-Spaces*

Linear models. Let $X=CP^n$ be the n-th complex projective space and G be a torus group acting on $\mathbb{C}^{n+1}$ via the representation $G\xrightarrow{\varphi}U(n+1)$ with the weight system $\Omega(\varphi)=\{w_j,k_j;1\leqslant j\leqslant s\}$. Then, there is an induced G-action on X and $H_G^*(X)\cong R[\xi]/\langle f(\xi)\rangle$ where $f(\xi)=(\xi-w_1)^{k_1}\cdots(\xi-w_s)^{k_s}$ and ξ is the transgression of the generator of $H^1(S)$ in the fibration $S^1\to S_G^{(2n+1)}\to CP_G^n$.

The above linear models demonstrate that not only all possibilities determined by the structure Theorem (VI.1) are geometrically realizable by linear models, but also the cohomological aspect of orbit structure of a *general* torus action on CCP^n is *the same as that of a suitable linear model.* Naturally, this kind of structure theorem motivates us to define a "weight system" as follows:

Definition. Let G be a compact connected Lie group, T be a maximal torus of G, and X be a CCP^n with a given G-action. Then the restricted T-action on X and $H_T^*(X)$ determine a set of weights with multiplicities $\{w_j,k_j,1\leqslant j\leqslant s\}$ as the roots of the structural equation $f(\xi)=0$ of $H_T^*(X)$. We shall call it *the weight system of* X and denote it by $\Omega(X)$. Notice that the structural equation $f(\xi)=0$ depends on the choice of the lifting ξ, and hence the above weight system $\Omega(X)$ is only defined up to a translation in the k-vector space $H^2(B_T)$.

The following are some general properties of the weight system that are direct consequences of Theorem (VI.1) and the definition:

Lemma (1.1). *Let G_1, G_2 be compact connected Lie groups, $h\colon G_1\to G_2$ be a Lie homomorphism, and T_1, T_2 be maximal tori of G_1, G_2 respectively such that*

$h(T_1) \subseteq T_2$. *Suppose* X *is a* G_2*-space of CCP-type, and* $h^*(X)$ *is the* G_1*-space structure on* X *induced by* h. *Then* $\Omega(h^*(X)) = h^*\Omega(X) = \{h^* w_j; w_j \in \Omega(X)\}$. *In particular, if* G_1 *is a subgroup of* G_2 *and* h *is the inclusion, we shall simply denote* $\Omega(h^*(X)) = h^*(\Omega(X))$ *by* $\Omega(X)|G_1$, *or rather* $\Omega(X)|T_1$.

Lemma (1.2). *Let* X *be a* CCP^n *with a given G-action and* $x \in X$ *be a given point. Assume that the maximal torus* T' *of* G_x^0 *is contained in the maximal torus* T *of* G. *Then there exist a suitable subcollection of weights in* $\Omega(X)$, *say* $\{w_{j1}, \ldots, w_{jl}\}$ *such that* T' *is given by the following equation:* $w_{j1} = \cdots = w_{jl}$. *Moreover,*

$$\Omega_{j1}|T' = \Omega(\mathrm{Ad}_G|T' - \mathrm{Ad}_{G_x}|T') + \Omega(S_x),$$

where Ω_{j1} *is the local weight system at* F^{j_1} *and* $\Omega(S_x)$ *is the local weight system of the* T'*-action on the slice,* S_x, *at* x.

Lemma (1.3). *It is always possible to choose suitable lifting* ξ *so that* $\Omega(X)$ *is invariant under the action of the Weyl group* $W(G)$ *on* $H^2(B_T)$.

Proofs the above lemmas are straightforward verifications.

(C) *The Geometric Weight System and the Principal Orbit Type*

Similar to the situation of transformation groups on acyclic manifolds discussed in the last chapter, the weight system, $\Omega(X)$, of a G-space X of CCP-type is a well-organized, simple set of invariants from which one can easily read off the orbit structure of the restricted T-action on X. Hence one may again expect that, to a large extent, the orbit structure of the original G-action on X can also be determined by $\Omega(X)$. The statements as well as the proofs of the results in this direction are similar to the case of acyclic manifolds treated in § 2 of last chapter. Let us mention here the following algorithm for computing the principal orbit type as an example of such results [cf. Th. (V.5)].

Theorem (VI.2). *Let* G *be a compact connected Lie group and* X *be a G-space of CCP-type. Let* T *be a maximal torus of* G, $\Omega(X)$ *be the weight system of* X *and* (H_X^0) *be the connected principal orbit type of* X. *Then, for each* $1 \leq j \leq s$, *there exist a sequence of non-zero weights* $\alpha_1, \ldots, \alpha_k$ *and a sequence of decreasing subtori* $T = S_0 \supsetneq S_1 \supsetneq \cdots \supsetneq S_k$ *satisfying the following recursive conditions with respect to the system of local weights at* $F^j(T, X)$, $\Omega_j(X)$, *and* S_k *is the maximal torus of a suitable* H_X^0, *namely,*

$S_0 = T; \alpha_1$: *an arbitrary non-zero weight in* $\{\Omega_j(X) - \Delta(G)\}$,

- -

$S_i = S_{i-1} \cap \alpha_i^\perp; \alpha_{i+1}$: *an arbitrary non-zero weight in* $\{\Omega_j(X)|S_i - \Delta(G)|S_i\}$,

- -

$S_k = S_{k-1} \cap \alpha_k^\perp$; $\{\Omega_j(X)|S_k - \Delta(G)|S_k\}$ *has no non-zero weights.*

Moreover, the root system of H_X^0, $\Delta(H_X^0)$, *is given by the following:*

$$\Delta(H_X^0) = \text{non-zero weights in } \{\Delta(G)|S_k - \Omega_j(X)|S_k\}.$$

Conversely, suppose that there exist such recursively defined sequence of weights $\alpha_1, \ldots, \alpha_k$ *and subtori leading from* T *to* S_k *(for at least one* $\Omega_j(X)$*). Then* S_k *is a* *maximal torus of a suitable principal isotropy subgroup* H_X^0 *with* $\Delta(H_X^0)$ = *non-zero weights in* $\{\Delta(G)|S_k - \Omega_j(X)|S_k\}$.

Corollary 1. *Suppose* X_1, X_2 *are two G-spaces of CCP-type and* $\Omega(X_1) \supseteq \Omega(X_2)$. *Then* $(H_{X_1}^0) \leqslant (H_{X_2}^0)$. *Hence, in particular,* $\Omega(X_1) = \Omega(X_2)$ *implies* $(H_{X_1}^0) = (H_{X_2}^0)$.

Corollary 2. *The conjugacy class of the above subtorus* S_k *(with respect to the action of Weyl group* $W(G)$*) is independent of the choice of* $\Omega_j(X)$ *and the choice of* α_i*'s.*

Proof of Theorem (VI.2). Let $q_j \in F^j$ be an arbitrary point of F^j and $G_j = G_{q_j}^0$, $G \supseteq G_j \supseteq T$. Then $\Omega_j(X) = \{\Delta(G) - \Delta(G_j)\} + \Omega(\mathcal{S}_j)$, where $\Omega_j(X)$ is the system of local weights at q_j and $\Omega(\mathcal{S}_j)$ is the weight system of the slice, $\mathcal{S}_j$, at q_j. It follows from the above equation that

$$\{\Omega_j(X)|S_i - \Delta(G)|S_i\} = \{\Omega(\mathcal{S}_j)|S_i - \Delta(G_j)|S_i\}$$

for any subtorus S_i of T. Hence, Theorem (VI.2) follows from Theorem (V.5) by applying it to the G_j-action on the slice $\mathcal{S}_j$. Notice that the principal orbits are everywhere dense and the principal isotropy subgroups of the G_j-action on $\mathcal{S}_j$ are also principal isotropy subgroups of the G-action on X.

(D) *Classification of Principal Orbit Types for Actions of Simple Compact Lie Groups on CCP-Spaces*

Observe that the principal orbits are population-wise dominate everywhere, and hence the *type* of principal orbits is a *geometric characteristic* of overriding importance. On the other hand, for spaces of a given type such as spaces of CCP-type considered in this section, the *possibilities of principal orbit type* are usually rather limited and their classification naturally leads to neat results which are rather useful in the study of other geometric characteristics of transformation groups on CCP-spaces. A prototype of such a classification was first successfully carried out for *differentiable* actions on acyclic manifolds in [H 5, II, III] and for topological actions on acyclic cohomology manifolds in [H 16] (cf. § 3 and § 4 of Ch. V). Following the general idea and method of [H 16] and equipped with the effective algorithm of Theorem (VI.2) as our basic tool, one can without difficulty though with a bit of work carry out the classification of principal orbit types for actions of compact connected Lie groups on spaces of CCP-type. However, we shall only include here the *basic case of actions of simple compact Lie groups* as an indication of the general pattern of final results of such a classification problem.

Theorem (VI.3). *Let* G *be a simple compact connected Lie group and* X *be a* G*-space of CCP-type. Suppose that the connected principal isotropy groups,* (H_X^0), *are non-trivial. Then there exists a complex linear representation* φ *with* $\Omega'(\varphi) = \Omega'(X)$ *(i.e., has the same system of non-zero weights) and the induced action of* φ *on its complex projective space has the same connected principal isotropy subgroups.*

Sketch of proof. The detailed proof of Theorem (VI.3) is, in principle, very much the same as that of Theorem (V.6) for the case of acyclic manifolds. For each given simple compact connected Lie group G, the orbits of the $W(G)$-action on the weight space $H^2(B_G)$ usually consist of rather large number of weights except a few special orbits. Let us take the case $G=\mathrm{SU}(r+1)=A_r$ as an example. The weight space $H^2(B_G)$ is usually expressed in terms of a basis $(\theta_1, \theta_2, \ldots, \theta_{r+1})$ with a relation $\theta_1+\theta_2+\cdots+\theta_{r+1}=0$, the Weyl group $W(G)$ acts as permutation group of the θ's, and $W(G)\{k\cdot\theta_1\}$, $W(G)\cdot\{k(\theta_1+\theta_2)\}$ and $W(G)\{k\cdot(\theta_1+\theta_2+\theta_3)\}$ are the few special orbits.

Since the weight system $\theta(X)$ is invariant under the action of $W(G)$, $\Omega(X)$ decomposes into the sum of orbits of $W(G)$. It is not difficult to show by Theorem (VI.2) that $H_X^0\neq\{\mathrm{id}\}$ implies $\Omega(X)$ consists of only those special orbits with a small number of weights. Fortunately, those special orbits with a small number of weights are in fact the weight system of low dimensional complex representations of G. Hence Theorem (VI.3) follows from Corollary 1 of Theorem (VI.2).

(E) *Low Dimensional Topological Representations of Compact Lie Groups*

As another testing result, let us mention the following theorem:

Theorem (VI.4). *Let X be a CCP^n with an effective action of $G=\mathrm{Spin}(k)$. Then*

$$n\geqslant\begin{cases}2^r & \text{if } k=(2r+1) \text{ is odd,}\\ 2^{(r-1)} & \text{if } k=2r\not\equiv 0 \pmod 4,\\ 2^{(r-1)}+r & \text{if } k=2r\equiv 0 \pmod 4)\end{cases}$$

and the above estimates are best possible.

Proof. The proof of Theorem (VI.4) is again very much the same as that of Theorem (V.9) for the case of acyclic manifolds. Since the T-action on X is effective, the weight system $\Omega(X)$ must consists of at least one weight

$$w=k_1\theta_1+\cdots+k_r\theta_r$$

with $k_i\equiv\frac{1}{2} \pmod 1$. On the other hand, suppose $\Omega(X)$ contains such a weight w, then $\Omega(X)\supseteq W(G)\cdot w$ which consists of at least 2^r weight if $k=(2r+1)$ is odd, and at least $2^{(r-1)}$ weight if $k=2r$ is even. Hence, in the case $k=(2r+1)$ is odd, $(n+1)=$number of weights in $\Omega(X)\geqslant 2^r$. But $(n+1)=2^r$ implies that $\Omega(X)=\{\frac{1}{2}(\pm\theta_1\pm\theta_2\pm\cdots\pm\theta_r)\}$ and it follows from Theorem (VI.1) that

$$\mathrm{Ker}(X)\supseteq\mathrm{Ker}\,\Omega(X)=\{a\in T \text{ satisfying } w_1=\cdots=w_{2^r}\}=\mathbb{Z}_2,$$

which is a contradiction to the assumption that $\mathrm{Ker}(X)=\{\mathrm{id}\}$. Therefore $(n+1)>2^r$, that is $n\geqslant 2^r$, and it is clearly best possible because the linear action of $\mathrm{Spin}(2r+1)$ on CP^{2^r} induced by the representation $\varphi=\Delta_r+1$ with $\Omega(\varphi)=\{\frac{1}{2}(\pm\theta_1\pm\cdots\pm\theta_r),0\}$ is effective. The proof of the case $k=2r\not\equiv 0 \pmod 4$

is the same as above, except that the number of weights in the orbit $W(G)\cdot\{\frac{1}{2}(\theta_1+\cdots+\theta_r)\}$ is $2^{(r-1)}$ instead of 2^r. In the case $k=2r\equiv 0 \pmod 4$, the center of $\mathrm{Spin}(k)$ is $\mathbb{Z}_2+\mathbb{Z}_2$ and hence, $\Omega(X)$ needs another orbit besides $W(G)\cdot\{\frac{1}{2}(\theta_1+\theta_2+\cdots+\theta_r)\}$ to make $\mathrm{Ker}\,\Omega(X)=\{\mathrm{id}\}$. The weight system $\Omega(X)$ with the smallest number of weights and $\mathrm{Ker}\,\Omega(X)=\{\mathrm{id}\}$ is the following:

$$\Omega(X)=\{W(G)\cdot\{\tfrac{1}{2}(\theta_1+\cdots+\theta_r)\},\ W(G)\cdot(\theta_1),0\}$$

or

$$\Omega(X)=\{W(G)\cdot\{\tfrac{1}{2}(\theta_1+\cdots+\theta_{r-1}-\theta_r)\},\ W(G)\cdot(\theta_1),0\}.$$

Hence $(n+1)\geqslant 2^{(r-1)}+2r+1$ or $n\geqslant 2^{(r-1)}+2r$, and the proof of Theorem (VI.4) is complete. □

(F) *G-Admissible Weight Systems for Spaces of CP-Type*

Definition. Let G be a compact connected Lie group and T be a maximal torus of G. A system of weights with multiplicity in $H^2(B_T)$, Ω, is said to be *G-admissible* for spaces of CP-type if there exists a G-action on a space X of CP-type such that $\Omega(X)=\Omega$.

One of the most fundamental problems in the study of actions of compact connected Lie groups on CCP-spaces is the determination of those G-admissible weight systems for spaces of CP-type. So far, there are only a few rudimentary preliminary results in this direction. Let us mention here the following simple result.

Theorem (VI.5). *If Ω_1, Ω_2 are G-admissible for spaces of CP-type and G is a semi-simple compact connected Lie group, then $\Omega_1+\Omega_2$ is also G-admissible for spaces of CP-type.*

Proof. Let X_1, X_2 be G-spaces of CP-type with $\Omega(X_1)=\Omega_1$ and $\Omega(X_2)=\Omega_2$. Let ξ_1, ξ_2 be the generators of $H^*(X_1)$ and $H^*(X_2)$ respectively and t be the generator of $H^*(B_{S^1})=k[t]$. Let $f_i\colon X_i\to B_{S^1}$ with $f_i^*(t)=\xi_i$ and $S^1\to Y_i\to X_i$ be the induced bundle of f_i from the universal S^1-bundle; $S^1\to E_{S^1}\to B_{S^1}$. It is easy to see that Y_1, Y_2 are respectively cohomology spheres. Since G is semi-simple and Y_i is an S^1-bundle over X_i, it follows from a theorem of Stewart [S16] that the G-actions on X_i can be lifted to G-actions on Y_i. Hence $S^1\times G$ acts on Y_1, Y_2 and consequently on the joint $Y_1\circ Y_2$. It is then easy to see that the induced G-action on $Y_1\circ Y_2/S^1$ gives a G-space of CP-type and $\Omega(Y_1\circ Y_2/S^1)=\Omega(X_1)+\Omega(X_2)=\Omega_1+\Omega_2$. Thus, $\Omega_1+\Omega_2$ is also G-admissible for spaces of CP-type.

§ 2. Transformation Groups on Cohomology Quaternionic Projective Spaces

In this section, we shall summarize the results of a joint paper with J. C. Su [H20] which takes those cohomology manifolds of *rational cohomology type of quater-*

nionic projective spaces as the testing spaces. Again, we shall call them cohomology quaternionic projective spaces (abbreviated as CQP-spaces) and we shall use the rational field k as coefficients of all cohomology algebras in this section except otherwise specified. Algebraically, there is only a minute difference in the degrees of the generators of their cohomology algebras, namely, degree 2 for CCP-spaces and degree 4 for CQP-spaces. However, such a minute difference not only makes the corresponding structure theorem, Theorem (VI.6), much harder to prove but also allows some interesting new features and examples to occur. The crucial central result of this section is the following structure theorem:

(A) *A Structural Splitting Theorem for Tori Actions on CQP-Spaces*

Theorem (VI.6). *Let G be a torus group and X be a CQP^n with a given effective G-action. Suppose that $\mathrm{rk}(G)\geqslant 3$ if the fixed point set F consists of $(n+1)$ isolated points, and $\mathrm{rk}(G)\geqslant 2$ otherwise. Then*

(i) *There exists a unique lifting ξ of the generator ξ_0 of $H^*(X)$ such that all roots of the structural equation $f(\xi)=0$ of $H_G^*(X)\cong R[\xi]/\langle f(\xi)\rangle$ are perfect squares, i.e.,*

$$f(\xi)=(\xi-w_1^2)^{k_1},\ldots,(\xi-w_s^2)^{k_s}, \quad w_j\in H^2(B_G).$$

(ii) *Correspondingly, there are exactly s connected components, $\{F^j, 1\leqslant j\leqslant s\}$, of the fixed point set F and $F^j\sim_k QP^{(k_j-1)}$ if $w_j=0$, $F^j\sim_k CP^{(k_j-1)}$ otherwise. (Hence, there is at most one component of QP-type.)*

(iii) *The system of local weights at F^j, $\Omega_j(X)$, are given as follows*

$$\Omega_j(X)=\{\pm w_i\pm w_j, k_i(i\neq j);\ \pm w_j,(k_j-1); 0,2(k_j-1)\}.$$

(iv) *The induced homomorphism $\iota_j^*: H_G^*(X)\to H_G^*(F^j)=R\otimes H^*(F^j)$ is given as follows:*

$$\iota_j^*(\xi)=\begin{cases}\eta_j^2+2w_j\eta_j+w_j^2 & \text{if } F^j \text{ is of } CP\text{-type},\\ \xi_j & \text{if } F^j \text{ is of } QP\text{-type}\end{cases}$$

where η_j (resp. ξ_j) is the generator of $H^*(F^j)$.

(v) *Let $Y=F^0(x)$ be the F^0-variety of x in $X, K=G_x^0$ and $(F\cap Y)=F^{j_1}+\cdots+F^{j_l}$. Then*

$$Y\sim\begin{Bmatrix}CP^m\\ QP^m\end{Bmatrix} \text{ and } K \text{ is determined by } \begin{cases}\pm w_{j1}|K=\cdots=\pm w_{jl}|K\neq 0,\\ w_{j1}|K=\cdots\quad =w_{jl}|K=0\end{cases}$$

with a suitable choice of signs in the top equations and $m=(k_{j1}+\cdots+k_{jl}-1)$. Conversely, all subtori K given by the above two type of equations are elements of $\mathcal{O}^0(X)$.

Examples of linear models. Let G be a torus group acting on $\mathbb{Q}^{(n+1)}$ via the quaternionic linear representation $\varphi: G\to T^{n+1}\subseteq \mathrm{Sp}(n+1)$ with $\Omega(\varphi)=\{w_j,k_j; 1\leqslant j\leqslant s\}$ as its weight system. Then there is an induced G-action on QP^n. Let ξ be the transgression of the generator of $H^3(S^3)$ in the fibrations: $S^3\to S_G^{(4n+3)}\to QP_G^n$.

Then $H^*_G(QP^n) \cong R[\xi]/\langle f(\xi)\rangle$ where $f(\xi)=(\xi-w_1^2)^{k_1}\dots(\xi-w_s^2)^{k_s}=0$ is the defining structural equation. Correspondingly, the fixed point set F consists of s connected components, $\{F^j, 1\leqslant j\leqslant s\}$, $F^j=QP^{(k_j-1)}$ if $w_j=0$ and $F^j=CP^{(k_j-1)}$ otherwise. The weight systems of the local representation of G at F^j are given by

$$\Omega_j=\{\pm w_i\pm w_j, k_i (i\neq j);\ \pm w_j, (k_j-1); 0; 2(k_j-1)\}.$$

The induced homomorphism $\iota_j^*: H^*_G(QP^n)\to H^*_G(F^j)=R\otimes H^*(F^j)$ is given by

$$\iota_j^*(\xi)=\begin{cases}\eta_j^2+2w_j\eta_j+w_j^2 & \text{if } F^j=CP^{(k_j-1)}\\ \xi_j & \text{if } F^j=QP^{(k_j-1)} \quad (\text{i.e., } w_j=0).\end{cases}$$

Remarks. The above linear models show that all the possibilities determined by Theorem (VI.6) can be geometrically realized by linear models on the one hand, and on the other hand, the structure Theorem (VI.6) proves that the cohomological aspect of orbit structure of an arbitrary G-space X of CQP-type ($\mathrm{rk}(G)\geqslant 2$ or 3) must be the same as that of a suitable linear model.

(2) There do exist examples of T^1-action on manifolds $X\sim_k QP^n$ (k: rational field) with two components of fixed point sets of QP-type [S 4]. Hence, the restriction $\mathrm{rk}(G)\geqslant 2$ of the above theorem is indeed necessary.

(3) *If one assume that* $X\sim_{\mathbb{Z}} QP^n$ rather than $X\sim_k QP^n$), then it was proved by Bredon [B 16] that any T^1-action on X can have *at most one component* of the fixed point set of QP-type. Hence, it is interesting to know whether the above theorem still holds if the assumption $\mathrm{rk}(G)\geqslant 2$ is replaced by $X\sim_{\mathbb{Z}} QP^n$.

(B) *Proof of Theorem (VI.6).* The beginning part of the proof of Theorem (VI.6) is identical to that of Theorem (VI.1). Again, it follows from the evenness of $H^*(X)$ and the fundamental fixed point Theorem (IV.1) that

$$H^*_G(X)\cong R[\xi]/\langle f(\xi)\rangle \quad (\text{where } \xi \text{ is an arbitrary lifting}),$$
$$f(\xi)=(\xi-\alpha_1)^{k_1}\dots(\xi-\alpha_s)^{k_s}, \quad \alpha_j\in H^4(B_G)$$

and correspondingly,

$$F^j\sim_k CP^{(k_j-1)} \quad \text{or} \quad QP^{(k_j-1)}.$$

The delicate part of the proof is to show that there exists a unique lifting ξ such that all the above roots $\{\alpha_j\}$ of the structural equation $f(\xi)=0$ become perfect squares. It is for this part of proof that we need the assumption $\mathrm{rk}(G)\geqslant 2$ or 3 and the following lemmas:

Lemma (2.1). *If F^j is of CP-type and η_j is a generator of $H^*(F^j)$, $\deg(\eta_j)=2$, then $\beta_j\neq 0$ in the following expression*

$$\iota_j^*(\xi)=\alpha_j+2\beta_j\eta_j+\gamma_j\eta_j^2 \quad (\iota_j: F^j\subseteq X).$$

Proof. If $\beta_j=0$, then it is easy to show that $H^*(F^j)$ will be generated by η_j^2, which contradicts the assumption that $H^*(F^j)$ is generated by η_j.

Lemma (2.2). *Let f_j be the fundamental cohomology class of F^j, i.e.,*

$$f_j=\begin{cases}(0,\dots,1,0,\dots,0) & \text{if } F^j \text{ is an isolated point,}\\ (0,\dots,0,\eta_j^{(k_j-1)},0\cdots 0) & \text{if } F^j\sim CP^{(k_j-1)},\quad \deg\eta_j=2,\\ (0,\dots,0,\xi_j^{(k_j-1)},0\cdots 0) & \text{if } F^j\sim QP^{(k_j-1)},\quad \deg\xi_j=4.\end{cases}$$

Then the ideal $I(f_j)=I_{f_j}(X,F)$ [cf. § 2—(c) of Ch. IV for its definition] is a principal ideal generated by the following element:

$$\begin{cases} a_j=\beta_j^{(k_j-1)}\cdot\prod_{i\neq j}(\alpha_i-\alpha_j)^{k_i}, & \text{if } F^j\sim CP^{(k_j-1)},\\ a_j=\prod_{i\neq j}(\alpha_i-\alpha_j)^{k_i}, & \text{otherwise.}\end{cases}$$

Proof. Follows readily from the same computation as that of Theorem (VI.1).

Lemma (2.3). *For each pair $1\leqslant i,j\leqslant s$, $(\alpha_i-\alpha_j)$ splits into two linear factors, say $(\alpha_i-\alpha_j)=w\cdot w'$, $w,w'\in H^2(B_G)$. Moreover,*

(i) *if w, w' are linearly independent, then F^i, F^j are both of CP-type and there are exactly two corank one F^0-varieties of CP-type connecting F^i and F^j with $w^\perp$ and $w'^\perp$ as their respective generic isotropy subgroups.*

(ii) *if w, w' are linearly dependent, then there is only one corank one F^0-varieties of QP-type connecting F^i, F^j and with $w^\perp=w'^\perp$ as its generic isotropy subgroup.*

(iii) *if F^j is of CP-type, then there is a corank one F^0-variety of QP-type containing F^j and with $\beta_j^\perp$ as its generic isotropy subgroup.*

Proof. It follows from Lemma (2.2) and Theorem (IV.10).

Lemma (2.4). *If there exists a corank one subtorus $w^\perp=K\subseteq G$ which acts trivially on the whole X, i.e., $F(K,X)=X$, then*

$$(\alpha_i-\alpha_j)\equiv 0 \pmod{w^2}\quad \textit{for all}\quad 1\leqslant i,j\leqslant s$$

and

$$\beta_j\equiv 0 \pmod{w}\quad \textit{for all}\quad f^j \textit{ of CP-type.}$$

Proof. By the above assumption, X itself is the only corank one F^0-variety, and hence, all linear factors of a_j in Lemma (2.2) are proportional to w, i.e., $(\alpha_i-\alpha_j)\equiv 0 \pmod{w^2}$ and $\beta_j\equiv 0 \pmod{w}$.

Lemma (2.5). *Suppose that the G-action on X is effective and* $\mathrm{rk}(G)\geqslant 2$. *If F^j is of CP-type and $\iota_j^*(\xi)=\alpha_j+2\beta_j\eta_j+\gamma_j\eta_j^2$, then $\gamma_j\neq 0$.*

Proof. Suppose the contrary that $\gamma_j=0$. Then we claim that

$$(\alpha_i-\alpha_j)\equiv 0 \pmod{\beta_j}\quad \text{for all}\quad 1\leqslant i\leqslant s.$$

By Lemma (2.3), there always exists corank one F^0-variety Y connecting F^i and F^j. If Y is of QP-type, then it follows from Lemma (2.4) applying to Y instead of X that $(\alpha_i-\alpha_j)\equiv 0 \pmod{\beta_j^2}$. If Y is of CP-type. then there exists $\eta\in H_G^*(Y)$ with $\imath_j^*(\eta)=\eta_j$, $\imath_i^*(\eta)=\eta_i+\delta$. Hence, $\imath^*(\xi)=\alpha_j+2\beta_j\eta$ and $\imath_i^*(\xi)=\imath_i^*\imath^*(\xi)=\alpha_j+2\beta_j(\eta_i+\delta)$; consequently, $\alpha_i=\alpha_j+2\beta_j\delta$, or $(\alpha_i-\alpha_j)=2\beta_j\delta\equiv 0 \pmod{\beta_j}$. Therefore. again by Lemma 3, the corank one F^0-variety with $\beta_j^\perp$ as its generic isotropy subgroup must be of QP-type and connecting all F^i to F^j, which must be the whole space X. Thus, $\beta_j^\perp$ acts trivially on X which contradicts the assumptions $\mathrm{rk}(G)\geqslant 2$ and the effectiveness of the G-action. This contradiction proves that $\gamma_j\neq 0$.

Now we are ready to prove Theorem (VI.6) itself. We shall divide the proof into three cases: (1) there is a component of CP-type, (2) there is a component of QP-type; (3) F consists of $(n+1)$ isolated points.

(1) *Suppose there is a component, say F^1, of CP-type.* First, it is not difficult to *choose suitable generators* ξ, η_1 of $H_G^*(X)$ and $H^*(F^1)$ respectively so that $\imath_1^*(\xi)=\beta_1^2+2\beta_1\eta_1+\eta_1^2$. Then, we shall show that the roots $\{\alpha_j\}$ of the defining equation, $f(\xi)=0$, with respect to the above chosen generator ξ of $H_G^*(X)$ must be *all perfect squares.* If the linear factors of $(\alpha_1-\alpha_j)=(\beta_1^2-\alpha_j)$ are linearly independent, then, by Lemma (2.3) there are two corank one F^0-varieties of CP-type connecting F^1 and F^j. Let Y be one of them and η be the generator of $H_G^*(Y)$ such that $\imath_1^*(\eta)=\eta_1$, $\imath_j^*(\eta)=\eta_j+\delta$. Then it is clear that $\imath^*(\xi)=\beta_1^2+2\beta_1\eta+\eta^2$. Hence

$$\begin{aligned}\imath_j^*(\xi)&=\imath_j^*\imath^*(\xi)=\imath_j^*(\beta_1^2+2\beta_1\eta+\eta^2)\\&=\beta_1^2+2\beta_1(\eta_j+\delta)+(\eta_j+\delta)^2=(\beta_1+\delta)^2+2(\beta_1+\delta)\eta_j+\eta_j^2,\end{aligned}$$

that is $\alpha_j=(\beta_1+\delta)^2=w_j^2$ and $\imath_j^*(\xi)=(w_j^2+2w_j\eta_j+\eta_j^2)$.

In case the linear factors of $(\alpha_1-\alpha_i)=(\beta_1^2-\alpha_i)$ are linearly dependent, then, again by Lemma 3, there is one F^0-variety Y of QP-type connecting F^1 and F^i. It follows from Lemma (2.4), applying to Y, that

$$(\alpha_1-\alpha_i)=(\beta_1^2-\alpha_i)\equiv 0 \pmod{w^2} \quad\text{and}\quad \beta_1\equiv 0 \pmod{w},$$

which obviously implies that $\alpha_i=a\cdot\beta_1^2$, $a\in k$ is a rational number. We claim that $\sqrt{a}$ is still rational, for otherwise, it is elementary to show that $(\alpha_i-\alpha_j)=(a\beta_1^2-w_j^2)$ is irreducible in $R=k[t_1,\dots,t_r]$ for any w_j linearly independent to β_1. Hence $\alpha_i=(\sqrt{a}\,\beta_1)^2=w_i^2$ is again a perfect square.

Next we shall show that $F^j\sim_k QP^{(k_j-1)}$ implies $w_j^2=\alpha_j=0$. Suppose F^j is of QP-type, then all corank one F^0-varieties containing F^j must also be of QP-type. Hence, it follows from Lemma (2.3) that (w_j-w_i) and (w_j+w_i) are linearly dependent for $1\leqslant i\leqslant s$, which clearly implies $w_j=0$.

(2) *Suppose there is a component, say F^1, of QP-type.* For simplicity, we may assume in this case that all components are either isolated points or of QP-type. Let us first choose the lifting ξ so that $\imath_1^*(\xi)=\xi_1$. Then it follows easily from Lemma (2.3) and (2.4) that

$$\alpha_j=a_jw_j^2 \quad\text{for}\quad 2\leqslant j\leqslant s$$

where the w_j are primitive in the sense that they are linear forms in $(t_1,\dots,t_r)$ with *relative prime* integral coefficients. Again, it is elementary to show that the splitting of $(\alpha_i-\alpha_j)=(a_i w_i^2-a_j w_j^2)$ for all $2\leqslant i, j\leqslant s$ implies that $\sqrt{a_i/a_j}$ are all rational for $2\leqslant i, j\leqslant s$. Hence, it is possible to adjust ξ by a rational factor so that all α_j become perfect squares. The rest of the proof is identical with the case (1).

(3) *F consists of $(n+1)$ isolated points and* $\mathrm{rk}(G)\geqslant 3$. (i) Suppose there exists a point F^1, such that the two linear factor of $(\alpha_1-\alpha_j)$ are *linearly dependent for each* $2\leqslant j\leqslant(n+1)$. Then, by changing the lifting ξ, we may assume $\alpha_1=0$ to start with. Hence

$$\alpha_j=a_j w_j^2 \quad \text{for} \quad 2\leqslant j\leqslant(n+1)$$

and the same proof as that of case (2) applies. [In a way, this is the case where F^1 should be considered as QP^0.]

(ii) Suppose that the two linear factors of $(\alpha_1-\alpha_2)$ are linearly independent. Let Y, Y' be the two corank one F^0-varieties connecting F^1 and F^2. It is possible to choose generators ξ, η of $H_G^*(X)$ and $H_G^*(Y)$ respectively so that $\iota^*(\xi)=\beta^2+2\beta\eta+\eta^2$ [notice that we assume $\mathrm{rk}(G)\geqslant 3$]. Let $\iota_1^*(\eta)=\delta_1$, $\iota_2^*(\eta)=\delta_2$. Then it is easy to see that

$$\begin{aligned}\alpha_1&=\iota_1^*\,\iota^*(\xi)=\iota_1^*(\beta^2+2\beta\eta+\eta^2)=(\beta+\delta_1)^2=w_1^2,\\ \alpha_2&=\iota_2^*\,\iota(\xi)=(\beta+\delta_2)^2=w_2^2.\end{aligned}$$

Let $K=(w_1-w_2)^\perp$. Then the proof of case (1) applied to the restricted K-action on X implies that

$$\alpha_j\equiv\tilde{w}_j^2 \bmod (w_1-w_2) \quad 3\leqslant j\leqslant(n+1)$$

for suitable $\tilde{w}_j\in H^2(B_G)$.

Applying the same argument to the other corank one F^0-variety Y' with $G_{Y'}=K'=(w_1+w_2)^\perp$, we have

$$\alpha_j\equiv\tilde{w}_j'^2 \bmod (w_1+w_2), \qquad 3\leqslant j\leqslant(n+1)$$

for suitable $\tilde{w}_j'\in H^2(B_G)$. Since (w_1-w_2) and (w_1+w_2) are linearly independent, it is easy to combine the above two congruences to show that there exist suitable $w_j\in H^2(B_G)$ and $a_j\in k$ such that

$$\alpha_j=w_j^2+a_j(w_1^2-w_2^2),$$

for each $3\leqslant j\leqslant(n+1)$. Suppose w_j, w_1, w_2 are linearly independent for a given $3\leqslant j\leqslant(n+1)$ [such j exists for $\mathrm{rk}(G)\geqslant 3$]. Then it follows from the splitting of both

$$(\alpha_1-\alpha_j)=w_1^2-w_j^2-a_j(w_1^2-w_2^2)$$

and

$$(\alpha_2-\alpha_j)=w_2^2-w_j^2-a_j(w_1^2-w_2^2)$$

into linear factors that a_j must be zero, i.e., $\alpha_j=w_j^2$. Finally, we shall show that α_i is a perfect square even if $\alpha_i=a w_1^2+b w_1 w_2+c w_2^2$. This is an easy consequence of the splitting of $(\alpha_i-\alpha_j)=(\alpha_i-w_j^2)$ for a suitable j with w_j, w_1, w_2 linearly independent. Thus we have proved that all α's are perfect sequences, i.e.,

$$f(\xi)=(\xi-w_1^2)(\xi-w_2^2)\cdot\dots\cdot(\xi-w_{n+1}^2).$$

Combine all the above cases and the proof of Theorem (VI.6) is complete.

Remark. The above weights $w_1,\dots,w_s\in H^2(B_G,k)$ are rational weights. However, by suitable modification of the lifting ξ with a rational factor, we may assume that $w_1,\dots,w_s\in H^2(B_G,\mathbb{Z})$ are all integral weights, and moreover, there are no common divisors among the coefficients of w_j. Later on, we shall always assume the system of weights of X to be such integral weights.

§ 3. Structure Theorems for Actions of $\mathbb{Z}_p$-Tori on $\mathbb{Z}_p$-Cohomology Projective Spaces (p Odd Primes)

In this section, we shall briefly discuss how to adapt the structure theorems of § 1, § 2 for torus group actions on CCP^n and CQP^n to the setting of $\mathbb{Z}_p$-tori actions on $\mathbb{Z}_p$-CCP^n and $\mathbb{Z}_p$-CQP^n, when p are odd primes. The $\mathbb{Z}_2$-case is rather special and will be treated in § 4. We shall use $\mathbb{Z}_p$-coefficients for all cohomology algebras in this section except specified to be otherwise. Let G be a $\mathbb{Z}_p$-torus of rank l, p be an odd prime. Then

$$H^*(B_G)=\Lambda_{\mathbb{Z}_p}[v_1,\dots,v_l]\otimes_{\mathbb{Z}_p}\mathbb{Z}_p[t_1,\dots,t_l]$$

where $\deg(v_j)=1$, $\deg(t_j)=2$ and $\beta_p(v_j)=t_j$ (β_p is the Bockstein operation). We shall denote the polynomial part of $H^*(B_G)$ by R_G, or simply by R when there is no danger of confusion. Suppose that X is a $\mathbb{Z}_p$-CCP^n (resp. $\mathbb{Z}_p$-CQP^n), i.e., $H^*(X)\cong\mathbb{Z}_p[\xi_0]/\langle\xi_0^{n+1}\rangle$, $\deg(\xi_0)=2$ (resp. 4), with a given G-action. Then the E_2-terms of the Serre spectral sequence of the fibration $X\to X_G\to B_G$ is

$$E_2=H^*(B_G)\otimes_{\mathbb{Z}_p}H^*(X)$$

and the generator, ξ_0, of $H^*(X)$ is transgressive. Since $\beta_p(\xi_0)=0$ and the Bockstein operation β_p commutes with transgression, the transgression of ξ_0, $\tau(\xi_0)$, must lie in $\mathrm{Ker}(\beta_p)$. In most cases, it is not difficult to show, $\tau(\xi_0)=0$ and hence $E_2=E_\infty$, $F\neq\phi$.

Definition. The quotient of $H_G^*(X;\mathbb{Z}_p)$ by the ideal generated by $\{v_j\}$ is called the *β_p-reduced equivariant cohomology* of X, and denoted by $\bar{H}_G^*(X)$ instead of $H_G^*(X)$ itself. For example, it follows from $E_2=E_\infty$ that $\bar{H}_G^*(X)=R[\xi]/\langle f(\xi)\rangle$ where ξ is an arbitrary lifting of the generator ξ_0 of $H^*(X)$ and $f(\xi)=\xi^{n+1}+c_1\xi^n+\dots+c_{n+1}=0$ is the defining structural equation. Then, almost identical proofs will give the following $\mathbb{Z}_p$-version of structure Theorem (VI.1) and (VI.6):

Theorem (VI.1$_p$). *Let G be a $\mathbb{Z}_p$-torus (p odd prime), and X be a $\mathbb{Z}_p$-CCP^n with a given G-action, $F \neq \phi$. Let ξ be an arbitrary lifting of the generator ξ_0 of $H^*(X)$ into $\bar{H}_G^*(X)$. Then $\bar{H}_G^*(X) \cong R[\xi]/\langle f(\xi)\rangle$ where*

$$f(\xi) = (\xi - w_1)^{k_1} \cdots (\xi - w_s)^{k_s}), \quad w_j \in \bar{H}^2(B_G)$$

splits into the product of linear factors. And correspondingly, the fixed point set $F(G,X)$ consists of s connected components $\{F^j, 1 \leqslant j \leqslant s\}$ such that

(i) $F^j \sim_{\mathbb{Z}_p} CP^{(k_j-1)}$ *and* $\iota_j^*: \bar{H}_G^*(X) \to \bar{H}_G^*(q_j) = R$, $q_j \in F^j$ *map ξ to w_j,*

(ii) *the system of local p-weights at F^j, Ω_j, is given by*

$$\Omega_j = \{\pm(w_j - w_i), k_i (i \neq j); 0, 2(k_j - 1)\},$$

(iii) *for a given point $x \in X$ with $F(x) \cap F(G,X) = F^{j1} + \cdots + F^{jl}$, the isotropy subgroup G_x is given by $w_{j1} = \cdots = w_{jl}$ and $F(x) \sim_p CP^m$, $m = (k_{j1} + \cdots + k_{jl} - 1)$.*

Theorem (VI.6$_p$). *Let G be a $\mathbb{Z}_p$-torus, p odd prime, and X be a $\mathbb{Z}_p$-CQP^n with a given effective G-action, $F \neq \phi$. Suppose that $\mathrm{rk}(G) \geqslant 3$ if the fixed point set F consists of $(n+1)$ isolated points, and $\mathrm{rk}(G) \geqslant 2$ otherwise. Then*

(i) *There exists a unique lifting ξ of the generator ξ_0 of $H^*(X)$ such that all roots of the structural equation $f(\xi) = 0$ of $\bar{H}_G^*(X) \cong R[\xi]/\langle f(\xi)\rangle$ are perfect squares, i.e.,*

$$f(\xi) = (\xi - w_1^2)^{k_1} \cdots (\xi - w_s^2)^{k_s}, \quad w_j \in \bar{H}^2(B_G).$$

(ii) *Correspondingly, there are exactly s connected components, $\{F^j, 1 \leqslant j \leqslant s\}$ of the fixed point set F and $F^j \sim_p QP^{(k_j-1)}$ if $w_j = 0$ and $F^j \sim_p CP^{(k_j-1)}$ otherwise. (Hence, there is at most one component of QP-type.)*

(iii) *The system of local p-weights at F^j, Ω_j, is given as follows*

$$\Omega_j = \{\pm w_i \pm w_j), k_i (i \neq j); \pm w_j, 2(k_j - 1); 0, 2(k_j - 1)\}.$$

(iv) *The induced homomorphism $\iota_j^*: \bar{H}_G^*(X) \to \bar{H}_G^*(F^j) = R \otimes H^*(F^j)$ is given as follows:*

$$i_j^*(\xi) = \begin{cases} \eta_j^2 + 2w_j\eta_j + w_j^2 & \text{if } F^j \text{ is of } CP\text{-type}, \\ \xi_j & \text{if } F^j \text{ is of } QP\text{-type} \end{cases}$$

where η_j (resp. ξ_j) is the generator of $H^*(F^j)$.

(v) *Let $Y = F(x)$ be the F-variety of x in X, $K = G_x$ and $F \cap Y = F^{j1} + \cdots + F^{jl}$. Then*

$$Y \sim_p \begin{Bmatrix} CP^m \\ QP^m \end{Bmatrix} \text{ and } K \text{ is given by } \begin{cases} \pm w_{j1}|K = \cdots = \pm w_{jl}|K \neq 0, \\ \quad w_{j1}|K = \cdots = \quad w_{jl}|K = 0 \end{cases}$$

with suitable choice of signs in the top equations and $m = (k_{j1} + \cdots + k_{jl} - 1)$. Conversely, all subtori K given by the above two types of equations are elements of $\mathcal{O}(X)$.

The proofs of the above $\mathbb{Z}_p$-version structure theorems are parallel to their respective characteristic zero versions. We leave them to the reader. As one of the simplest consequence the bijection between distinct roots of the structural equation $f(\xi)=0$ and the components of the fixed point set, one has the following estimate of the number of components:

Corollary (3.1). *If X is a $\mathbb{Z}_p$-CCP^n (resp. $\mathbb{Z}_p$-CQP^n) then the number of connected components, s, of the fixed point set is at most $p^{\mathrm{rk}(G)}$ (resp. $\frac{1}{2}p^{\mathrm{rk}(G)}$).*

Proof. Since the connected components $\{F^j\}$ are indexed by distinct roots w_j (resp. $\{w_j^2\}$) in the case $X \sim_p CP^n$ (resp. $X \sim_p QP^n$), the number s is obviously bounded by the number of available elements in $\bar{H}^2(B_G)$ (resp. elements of the form w^2 in $\bar{H}^4(B_G)$). Hence $s \leqslant p^{\mathrm{rk}(G)}$ (resp. $s \leqslant \frac{1}{2}p^{\mathrm{rk}(G)}$). □

Remarks. The above corollary is essentially included in Theorem (3.1) and Theorem (4.4) of [B 16]. However, the above proof provides a better understanding and is more straightforward.

Similar to the case of transformations of compact connected Lie groups on CCP-spaces, one may again apply the above $\mathbb{Z}_p$-version structure theorems to maximal $\mathbb{Z}_p$-tori to set up a system of p-weights for compact transformation groups on $\mathbb{Z}_p$-CCP-spaces and $\mathbb{Z}_p$-CQP-spaces as follows:

Definition. Let G be a compact Lie group and X be a $\mathbb{Z}_p$-cohomology projective space with a given G-action. Suppose $\{H_i, 1 \leqslant i \leqslant h\}$ is a complete representative, one from each of the h conjugacy classes of maximal $\mathbb{Z}_p$-tori of G. Then, applying the above structure theorems to each of the restricted H_i-action on X, $1 \leqslant i \leqslant h$, one obtains a system of p-weights, $\Omega(X|H_i)$, for each $1 \leqslant i \leqslant h$. The totality of h such systems $\{\Omega(X|H_i); 1 \leqslant i \leqslant h\}$ is called the *p-weight system* of the G-space X.

Geometrically, the (connected) weight system defined in terms of the restriction to a maximal torus T, $\Omega(X|T)$, detects the connected isotropy subgroups, while the p-weight system defined in terms of the restricted actions of h representative maximal $\mathbb{Z}_p$-tori, i.e., $\Omega_p(X) = \{\Omega(X|H_i), 1 \leqslant i \leqslant h\}$ detects the p-component of the isotropy subgroups. In fact, in order to take the p-primary components of isotropy subgroups into full account, it is necessary and useful to introduce secondary p-weight system as follows:

For a given maximal $\mathbb{Z}_p$-torus H_1 of G, let $N(H_1)$ be the normalizor of H_1 in G and $W(H_1) = N(H_1)/H_1$. For a given G-space X, there is an induced action of $W(H_1)$ on $F(H_1, X)$. In case X is a $\mathbb{Z}_p$-cohomology projective space and $\bar{H}_1$ is a maximal $\mathbb{Z}_p$-torus of $W(H_1)$, then $F(G, X)$ is again a sum of $\mathbb{Z}_p$-cohomology projective spaces with the restricted $\bar{H}_1$-action, and it follows from the structure theorems that such an H_1-action on $F(G, X)$ again defines a set of p-weight system which we shall call it the *secondary p-weight system of the G-space X with respect to the pair* $(H_1, \bar{H}_1)$. Let us take the simple case $G = \mathbb{Z}_{p^2} \oplus \cdots \oplus \mathbb{Z}_{p^2} = \mathbb{Z}_{p^2}^l$ as an example. Then there is a *unique* maximal $\mathbb{Z}_p$-torus, $H = \mathbb{Z}_p^l$, and $G/H \cong \mathbb{Z}_p^l$ is itself a $\mathbb{Z}_p$-torus. Suppose X is a given G-space of $\mathbb{Z}_p$-cohomology type of projective space. Then $F(H, X)$ is the sum of $\mathbb{Z}_p$-cohomology projective spaces, say $F(H, X) = F^1 + \cdots + F^s$ which are indexed by the distinct roots $\alpha_1, \ldots, \alpha_s$ of the structural equation $f(\xi)=0$. Since the induced action (via conjugation) of $\bar{H} = G/H$

on H and hence also on $\bar{H}^*(B_H)$ is trivial, each connected component, F^j, is invariant under $\bar{H}$. Hence, applying the structure theorems to the $\bar{H}$-action on each component F^j, one obtains altogether s secondary p-weight systems, $\Omega(F^j)$, $1 \leqslant j \leqslant s$.

Next, let us consider the case of a general finite group G. For a given prime p, there is a unique conjugacy class of maximal p-primary subgroups, the Sylow p-groups of G. Observe that any p-primary finite group is solvable. Suppose X is a G-space of *$\mathbb{Z}$-cohomology type* of projective space. Then X is a $\mathbb{Z}_p$-cohomology projective space for each prime p, and the $\mathbb{Z}_p$-version structure theorems enable us to define a hierarchy of p-weight systems in terms of the restriction of the G-action on X to a Sylow p-group, G_p, of G.

§ 4. Structure Theorems for Actions of $\mathbb{Z}_2$-Tori on $\mathbb{Z}_2$-Cohomology Projective Spaces

In this section, we consider the special $\mathbb{Z}_2$-case. The main results are the following structure theorems of [H 19] for actions of $\mathbb{Z}_2$-tori on spaces of $\mathbb{Z}_2$-cohomology type of real, complex and quaternionic projective spaces respectively.

Theorem (VI.7). *Let G be a $\mathbb{Z}_2$-torus and $X \sim_2 RP^n$ with a given G-action, $F(G,X) \neq \emptyset$ (non-empty). Then, the $\mathbb{Z}_2$-equivariant cohomology of X is isomorphic to $R[\xi]/\langle f(\xi)\rangle$, $\deg(\xi)=1$ as an R-algebra, where*

$$f(\xi)=(\xi+w_1)^{k_1}\cdots(\xi+w_s)^{k_s}$$

is the monic splitting polynomial of degree $(n+1)$ with $w_j \in H^1(B_G, \mathbb{Z}_2)$ as distinct roots. Correspondingly, the fixed point set F consists of s connected components $\{F^j, 1 \leqslant j \leqslant s\}$, $s \leqslant 2^{\mathrm{rk}(G)}$, such that

(i) *$F^j \sim_2 RP^{(k_j-1)}$ and $\iota_j^*: H_G^*(X) \to H_G^*(q_j)=R$, $q_j \in F^j$, maps ξ to w_j,*

(ii) *the system of 2-weights at F^j, Ω_j, is given by*

$$\Omega_j=\{(w_i+w_j),\, k_i (i \neq j);\, 0,\, (k_j-1)\},$$

(iii) *for a given point $x \in X$ with $F(x) \cap F(G,X)=F^{j1}+\cdots+F^{jl}$, the isotropy subgroup G_x is given by $w_{j1}=\cdots=w_{jl}$ and $F(x) \sim_2 RP^m$, $m=(k_{j1}+\cdots+k_{jl}-1)$.*

Theorem (VI.8). *Let G be a $\mathbb{Z}_2$-torus and X be G-space of $\mathbb{Z}_2$-cohomology type of CP^n, $F(G,X) \neq \emptyset$ (non-empty). Then $H_G^*(X) \cong R[\eta]/\langle f(\eta)\rangle$, where η is a suitable lifting of the generator of η_0 of $H^*(X)$ such that $Sq^1(\eta)=\beta \cdot \eta$ and $f(\eta)=(\eta+\alpha_1)^{k_1} \ldots (\eta+\alpha_s)^{k_s}$ is a splitting polynomial with $\alpha_j \in H^2(B_G)$ as distinct roots. Moreover, there are the following two cases according to $\beta=0$ or $\beta \neq 0$:*

(a) *In case $\beta=0$. Then all connected components of $F(G,X)$ are of CP-type, i.e., $F^j \sim_2 CP^{(k_j-1)}$ and all roots α_j are perfect squares, say $\alpha_j=w_j^2$. And furthermore, the system of local weights at F^j, Ω_j, is given by*

$$\Omega_j=\{(w_i+w_j),\, 2k_i (i \neq j);\, 0,\, 2(k_j-1)\}$$

and, for a given point $x \in X$ *with* $F(x) \cap F(G,X) = F^{j1} + \cdots + F^{jl}$, *the isotropy group* G_x *is given by* $w_{j1} = \cdots = w_{jl}$, $F(x) \sim_2 CP^m$, $m = (k_{j1} + \cdots + k_{jl} - 1)$.

(b) *In case* $\beta \neq 0$. *Then all connected components of* $F(G,X)$ *are of* RP*-type, i.e.,* $F^j \sim_2 RP^{(k_j-1)}$ *and the roots* α_j *are of the form* $\beta w_j + w_j^2$. *Furthermore, the system of local weights at* F^j, Ω_j, *is given by*

$$\Omega_j = \{(w_i + w_j), (w_i + w_j + \beta), k_i (i \neq j); \beta, 0, (k_j - 1)\}$$

and $F(x)$ *is of* CP*-type if and only if* $\beta | G_x = 0$.

Theorem (VI.9). *Let* G *be a* $\mathbb{Z}_2$*-torus and* X *be a* G*-space of* $\mathbb{Z}_2$*-cohomology type of* QP^n, $F(G,X) \neq \emptyset$ *(non-empty). Then, there exists a lifting* ζ *of generator* ζ_0 *of* $H^*(X)$ *into* $H_G^*(X)$ *such that* $Sq^1(\zeta) = 0$, $Sq^2(\zeta) = \gamma \cdot \zeta$ *and* $H_G^*(X) \cong R[\zeta]/\langle f(\zeta) \rangle$, *where*

$$f(\zeta) = (\zeta + \alpha_1)^{k_1} \dots (\zeta + \alpha_s)^{k_s}$$

is a splitting polynomial with $\alpha_j \in H^4(B_G)$ *as distinct roots. Moreover, there are the following three cases:*

(a) $\mathrm{Sq}^2(\zeta) = \gamma \cdot \zeta = 0$. *Then all connected components of* $F(G,X)$ *are of* QP*-type, i.e.,* $F^j \sim_2 QP^{(k_j-1)}$ *and the roots* $\alpha_j = w_j^4$ *for suitable* $w_j \in H^1(B_G)$. *And furthermore, the system of local weights at* F^j, Ω_j, *is given by*

$$\Omega_j = \{(w_i + w_j), 4k_i (i \neq j); 0, 4(k_j - 1)\}$$

and, for a given point $x \in X$ *with* $F(x) \cap F(G,X) = F^{j1} + \cdots + F^{jl}$, *the isotropy group* G_x *is given by* $w_{j1} = \cdots = w_{jl}$, $F(x) \sim_2 QP^m$, $m = (k_{j1} + \cdots + k_{jl} - 1)$.

(b) $\mathrm{Sq}^2(\zeta) = \gamma \cdot \zeta \neq 0$ *and* $\mathrm{Sq}^1(\gamma) = 0$: *Then all the connected components of* $F(G,X)$ *are of* CP*-type, i.e.,* $F^j \sim_2 CP^{(k_j-1)}$ *and the roots* α_j *are of the form* $\alpha_j = v^2 w_j^2 + w_j^4$, *where* $\gamma = v^2$, $w_j, v \in H^1(B_G)$. *Furthermore, the system of local weights at* F^j, Ω_j, *is given by*

$$\Omega_j = \{(w_i + w_j), (w_i + w_j + v), 2k_i (i \neq j); v, 0, 2(k_j - 1)\}$$

and $F(x)$ *is of* CP*-type (resp.* QP*-type) if* $v | G_x \neq 0$ *(resp.* $v | G_x = 0$*).*

(c) $\mathrm{Sq}^2(\zeta) = \gamma \cdot \zeta \neq 0$ *and* $\mathrm{Sq}^1(\gamma) \neq 0$; *then all connected components of* $F(G,X)$ *are of* RP*-type, i.e.,* $F^j \sim_2 RP^{(k_j-1)}$ *and there exist* $w_j, v_1, v_2 \in H^1(B_G)$ *such that*

$$\gamma = v_1^2 + v_1 v_2 + v_2^2, \qquad \alpha_j = w_j (w_j + v_1)(w_j + v_2)(w_j + v_1 + v_2)$$

and

$$\iota_j^*(\zeta) = \xi_j^4 + \gamma \xi_j^2 + (\mathrm{Sq}^1 \gamma) \xi_j + \alpha_j, \qquad \iota_j^*: H_G^*(X) \to H_G^*(F^j).$$

And furthermore, the system of local weights at F^j, Ω_j, *is given by*

$$\Omega_j = \{(w_i + w_j), (v_1 + w_i + w_j), (v_2 + w_i + w_j), (v_1 + v_2 + w_i + w_j), k_i (i \neq j); \\ v_1, v_2, (v_1 + v_2), 0, (k_j - 1)\},$$

and

$$F(x) \text{ is of } \begin{cases} RP\text{-type if } v_1|G_x \text{ and } v_2|G_x \text{ are linearly independent,} \\ CP\text{-type if } v_1|G_x \text{ and } v_2|G_x \text{ are linearly dependent but not all zero,} \\ QP\text{-type if } v_1|G_x = v_2|G_x = 0. \end{cases}$$

The proof of Theorem (VI.7) is parallel to the proof of Theorem (VI.1). However, the proofs of Theorem (VI.8) and (VI.9) do have a new feature involving the Steenrod square operations.

(A) *Examples of Linear Models*

Example 1. Let G be a $\mathbb{Z}_2$-torus action on $\mathbb{R}^{n+1}$ via a linear representation $\varphi: G \to 0(n+1)$. Suppose the system of $\mathbb{Z}_2$-weights of φ is $\Omega(\varphi) = \{w_j, k_j; 1 \leqslant j \leqslant s\}$ and $X = RP^n$ with the induced G-action of the above G-action on $\mathbb{R}^{(n+1)}$. Let $\xi \in H^1(X_G)$ be the transgression of the fibration $\mathbb{Z}_2 \to S^n_G \to RP^n_G$. Then it is not difficult to see that $H^*_G(X) \cong R[\xi]/\langle f(\xi)\rangle$ where $f(\xi) = (\xi + w_1)^{k_1} \ldots (\xi + w_s)^{k_s}$.

Example 2a. Let G be a $\mathbb{Z}_2$-torus acting on $\mathbb{C}^{n+1}$ via a complex linear representation $\varphi: G \to U(n+1)$. Suppose the system of $\mathbb{Z}_2$-weights of $\varphi: G \to \mathbb{Z}_2^{(n+1)} \subseteq U(n+1)$ is $\Omega(\varphi) = \{w_j, k_j; 1 \leqslant j \leqslant s\}$ and $X = CP^n$ with the induced G-action of the above G-action on $\mathbb{C}^{(n+1)}$. Then, it is easy to show that $H^*_G(X) \cong R[\eta]/\langle f(\eta)\rangle$, where η is the transgression of the fibration $S^1 \to S^{2n+1}_G \to CP^n_G$ and the defining equation $f(\eta) = (\eta + w_1^2)^{k_1} \ldots (\eta + w_s^2)^{k_s}$.

Example 2b. Let $G = \mathbb{Z}_2 \times G'$ be a $\mathbb{Z}_2$-torus acting on $\mathbb{C}^{n+1} = \mathbb{C} \otimes \mathbb{R}^{n+1}$ via a real linear representation $\varphi = \beta \otimes \varphi'$ where β is the conjugation of $\mathbb{C}$ and φ' is a real linear representation $G' \to \mathbb{Z}_2^{(n+1)} \subseteq 0(n+1)$ with $\Omega(\varphi') = \{w_j, k_j; 1 \leqslant j \leqslant s\}$. Then, there is an induced G-action on CP^n and it is not difficult to show that $H^*_G(X) \cong R[\eta]/\langle f(\eta)\rangle$ where

$$f(\eta) = (\eta + \beta w_1 + w_1^2)^{k_1} \ldots (\eta + \beta w_s + w_s^2)^{k_s}, \qquad Sq^1(\eta) = \beta \cdot \eta$$

and η is the transgression of the fibration $S^1 \to S^{2n+1}_G \to CP^n_G$, β is the $\mathbb{Z}_2$-weight corresponding to the conjugation of $\mathbb{C}$.

Example 3a. Let G be a $\mathbb{Z}_2$-torus acting on $\mathbb{Q}^{(n+1)}$ via a quaternionic representation $\varphi: G \to \mathbb{Z}_2^{(n+1)} \subseteq \mathrm{Sp}(n+1)$ with the system of $\mathbb{Z}_2$-weights $\Omega(\varphi) = \{w_j, k_j; 1 \leqslant j \leqslant s\}$. Let ζ be the transgression of the fibration $S^3 \to S^{4n+3}_G \to QP^n_G$. Then, $H^*_G(X) \cong R[\zeta]/\langle f(\zeta)\rangle$ where

$$f(\zeta) = (\zeta + w_1^4)^{k_1} \ldots (\zeta + w_s^4)^{k_s}.$$

Example 3b. Let $G = \mathbb{Z}_2 \times G'$ be a $\mathbb{Z}_2$-torus acting on $\mathbb{Q}^{(n+1)} = \mathbb{Q} \otimes \mathbb{R}^{(n+1)}$ via the real representation $\varphi = v \otimes \varphi'$, where v is the $\mathbb{Z}_2$-automorphism of $\mathbb{Q}$ that changes the signs of j, k and φ' is the real representation of $G' \to \mathbb{Z}_2^{(n+1)} \subseteq 0(n+1)$ with $\Omega(\varphi) = \{w_j, k_j; 1 \leqslant j \leqslant s\}$. Then there is an induced G-action on $X = QP^n$ and it is not difficult to show that $H^*_G(X) \cong R[\zeta]/\langle f(\zeta)\rangle$ where ζ is the transgression of the fibration $S^3 \to S^{4n+3}_G \to QP^n_G$ and $\mathrm{Sq}^2(\zeta) = v^2 \cdot \zeta$,

$$f(\zeta)=(\zeta+v^2 w_1^2+w_1^4)^{k_1}\dots(\zeta+v^2 w_s^2+w_s^4)^{k_s}.$$

Example 3c: Notice that the effective group of isometry on the symmetric space QP^n is $\mathrm{Sp}(n+1)/\{\pm 1\}$ and $\mathrm{SO}(3)=\mathrm{Sp}(1)/\{\pm 1\}$ sits in $\mathrm{Sp}(n+1)/\{\pm 1\}$ as diagonal unit quaternions. Let G be a $\mathbb{Z}_2$-tori of $\mathrm{Sp}(n+1)/\{\pm 1\}$ containing a maximal $\mathbb{Z}_2$-tori of $\mathrm{SO}(3)$. Then the restricted G-action on QP^n is of the type of case 3.

The above "linear models" not only demonstrate that all the possibilities in the statements of the above structure theorems are *geometrically realizable* but also show that the *cohomological aspect of orbit structures of general G-actions on cohomology projective spaces are the same as that of suitable linear models.*

(B) *Proofs of the Structure Theorems*

Since the proof of Theorem (VI.7) is parallel to the proof of Theorem (VI.1), and the proof of Theorem (VI.8) is similar to but simpler than the proof of Theorem (VI.9), we shall only prove Theorem (VI.9) in the following and leave the proofs of Theorems (VI.7) and (VI.8) to the reader.

Proof of Theorem (VI.9). We choose the lifting ζ of the degree 4 generator ζ_0 of $H^*(X)$ into $H_G^*(X)$ so that α_1 = the restriction of ζ to $H_G^*(q_1)=0$. Then it is easy to see that $\mathrm{Sq}^1(\zeta)=\beta\zeta$ and $\mathrm{Sq}^2(\zeta)=\gamma\cdot\zeta$, $\beta\in H^1(B_G, \gamma\in H^2(B_G)$. Namely $\mathrm{Sq}^1(\zeta)$ and $\mathrm{Sq}^2(\zeta)$ restrict to zero in $H_G^*(q_1)$.

(i) We claim that $\mathrm{Sq}^1(\zeta)=\beta\cdot\zeta=0$. Suppose the contrary that $\mathrm{Sq}^1(\zeta)=\beta\cdot\zeta\neq 0$, i.e., $\beta\neq 0\in H^1(B_G)$. Then there is a rank one $\mathbb{Z}_2$-subtorus K of G such that $\beta|K\neq 0$. Let t be the generator of $H^*(B_K)$ and $\bar{\zeta}$ be the image of ζ in $H_K^*(X)$. By the naturality of Sq^1, we have $\mathrm{Sq}^1(\bar{\zeta})=t\cdot\bar{\zeta}\neq 0$. We claim that $F(K,X)$ must be connected. For otherwise, $H_K^*(X)\cong\mathbb{Z}_2[t][\bar{\zeta}]/\langle\bar{f}(\bar{\zeta})\rangle$ and the defining equation $\bar{f}(\bar{\zeta})=0$ *must have more than one distinct root* $\{\bar{\alpha}_j\in H^4(B_K)\}$. However, $H^4(B_K)$ consists of only two elements, namely, 0 and t^4; we see that $\bar{\alpha}_1=0$, and $\bar{\alpha}_2=t^4$ must be the two distinct roots of $\bar{f}(\bar{\zeta})=0$. Then

$$0=\mathrm{Sq}^1(\bar{\alpha}_2)=\mathrm{Sq}^1(\bar{\zeta}|H_K^*(\bar{q}_2)=\mathrm{Sq}^1(\zeta)|H_K^*(\bar{q}_2)=t\bar{\zeta}|H_K^*(\bar{q}_2)=t^5$$

which is clearly a contradiction. Hence $F(K,X)$ must be connected and we may assume that $\bar{\alpha}_1=0$. Then there are the following three possibilities, namely, $F(K,X)\sim_2 QP^n$, or CP^n, or RP^n. In case $F(K,X)\sim_2 QP^n$ or CP^n, it is easy to show that $\imath^*\mathrm{Sq}^1(\bar{\zeta})=\mathrm{Sq}^1(\imath^*\bar{\zeta})=0$ and hence, by the injectivity of $\imath^*: H_K^*(X)\to H_K^*(F(K,X))$, $\mathrm{Sq}^1(\bar{\zeta})=0$ which contradicts the assumption. In case $F(K,X)\sim_2 RP^n$, one has

$$\imath^*(\bar{\zeta})=t^3\bar{\xi}_1+a_2\bar{\xi}_1^2+a_3\bar{\xi}_1^3+a_4\bar{\xi}_1^4$$

where $\bar{\xi}_1$ is the generator of $H^*(F(K,X))$. Then

$$\begin{array}{l} t\cdot\imath^*(\zeta)=t^4\bar{\xi}_1+t\cdot a_2\bar{\xi}_1^2+ta_3\bar{\xi}_1^3+ta_4\bar{\xi}_1^4 \\ \Vert \\ \mathrm{Sq}^1(\imath^*(\bar{\zeta}))=t^4\bar{\xi}_1+(t^3+\mathrm{Sq}^1 a_2)\bar{\xi}_1^2+\cdots \end{array}$$

and the fact that $\mathrm{Sq}^1(a_2)=0$ no matter $a_2=0$ or t^2 implies that $a_2=t^2$. Notice that $\mathrm{Sq}^2(\bar{\zeta})=0$ or $t^2\bar{\zeta}$, which implies that $\mathrm{Sq}^2(\iota^*(\bar{\zeta}))=0$ or $t^2(\iota^*(\bar{\zeta}))$. However, $\mathrm{Sq}^2(\iota^*(\bar{\zeta}))=\mathrm{Sq}^2(t^3\bar{\xi}_1+t^2\bar{\xi}_1^2+\cdots)=(t^5\bar{\xi}_1+0\cdot\bar{\xi}_1^2+\cdots)$ which is neither 0 nor $t^2(\iota^*(\bar{\zeta}))=(t^5\bar{\xi}_1+t^4\xi_1^2+\cdots)$, hence, a contradiction. All the above contradictions prove that the assumption $\mathrm{Sq}^1(\zeta)=\beta\cdot\zeta\neq 0$ is impossible, that is, $\mathrm{Sq}^1(\zeta)=0$.

(ii) *The case* $\mathrm{Sq}^2(\zeta)=\gamma\cdot\zeta=0$. We claim that there are no components of either RP-type or CP-type. In fact, it is easy to show that the existence of a component of RP-type implies $\mathrm{Sq}^1(\zeta)\neq 0$ and the existence of a component of CP-type implies $\mathrm{Sq}^2(\zeta)\neq 0$. Moreover $\mathrm{Sq}^1(\zeta)=0$, $\mathrm{Sq}^2(\zeta)=0$ imply that

$$Sq^1(\alpha_j)=0 \quad \text{and} \quad Sq^2(\alpha_j)=0 \quad \text{for all} \quad 1\leqslant j\leqslant s,$$

and it is not difficult to show that such degree 4 elements must be of the form $\alpha_j=w_j^4$, $w_j\in H^1(B_G)$. Therefore

$$\iota^*(\zeta)=(\zeta_1+w_1^4,\ \zeta_2+w_2^4,\ldots,\zeta_s+w_s^4)\in H_G^4(F),$$

and the same computation as in the proof of Theorem (VI.1) will show that

$$I_X(f_j)=\textstyle\prod_{i\neq j}(w_i^4+w_j^4)^{k_i}=\prod_{i\neq j}(w_i+w_j)^{4k_i}.$$

The rest of the proof of this case is again the same as that of Theorem (VI.1).

(C) *The case* $\mathrm{Sq}^2(\zeta)=\gamma\cdot\zeta\neq 0$ and $\mathrm{Sq}^1(\gamma)=0$: In this case, it is easy to show that all the connected components are of CP-type. Let η_j be the generator of $H^*(F^j)$, ι_j^* be the restriction homomorphism of $H_G^*(X)$ to $H_G^*(F^j)$, and $\iota_j^*(\zeta)=(a_j\eta_j^2+b_j\eta_j+\alpha_j)$. Then

$$\mathrm{Sq}^2(\zeta)=\gamma\cdot\zeta \quad \text{implies} \quad a_j=1,\ b_j=\gamma \quad \text{and} \quad \mathrm{Sq}^2\alpha_j=\gamma\cdot\alpha_j,$$

Hence $\gamma=v^2$ is a perfect square. Now, let us choose basis in $H^1(B_G)$ so that v is the first base, and express α_j as polynomial in v with coefficients in terms of the other "variables". Then it is not difficult to show that $\mathrm{Sq}^1(\alpha_j)=0$ and $\mathrm{Sq}^2(\alpha_j)=v^2\cdot\alpha_j$ imply that $\alpha_j=v^2\cdot w_j^2+w_j^4$ for a suitable $w_j\in H^1(B_G)$. Therefore,

$$\iota^*(\zeta)=(\eta_1^2+v^2\eta_1+v^2w_1^2+w_1^4,\ldots,\eta_s^2+v^2\eta_s+v^2w_s^2+w_s^4)\in H_G^4(F)$$

and the rest of the proof of this case follows easily from a similar computation as that of Theorem (VI.1).

In the case $\mathrm{Sq}^2(\zeta)=\gamma\cdot\zeta\neq 0$ *and* $\mathrm{Sq}^1(\gamma)=\delta\neq 0$, it is not difficult to show that all connected components of $F(G,X)$ are of RP-type. Let ξ_j be the generator of $H^*(F^j)$, ι_j^* be the restriction homomorphism of $H_G^*(X)$ to $H_G^*(F^j)$. Then, it follows from the following equations:

$$\mathrm{Sq}^1(\iota_j^*\zeta)=\iota_j^*(\mathrm{Sq}^1\zeta)=0 \quad \text{and} \quad \mathrm{Sq}^2(\iota_j^*\zeta)=\iota_j^*(\mathrm{Sq}^2\zeta)=\gamma\cdot(\iota_j^*\zeta)$$

that

$$\iota_j^*\zeta=\xi_j^4+\gamma\cdot\xi_j^2+\delta\xi_j+\alpha_j;$$

$$\mathrm{Sq}^1\delta=\mathrm{Sq}^1\alpha_j=0, \quad \mathrm{Sq}^2\delta=\gamma\delta, \quad \mathrm{Sq}^2\alpha_j=\gamma\alpha_j.$$

Let $f_j=(0,\ldots,0,\zeta_j^{(k_j-1)},0,\ldots,0)$ be the fundamental cohomology class of F_j. Then, simple computation will show that the ideal $I_X(f_j)$ is generated by $a(f_j) =\delta^{(k_j-1)}\cdot\prod_{i\neq j}(\alpha_i+\alpha_j)$ (cf. the proof of Theorem 1). For simplicity, we may assume without loss of generality that $\alpha_1=0$.

(1) *Suppose that at least one* $k_j>1$:

Then it follows from Theorem B that δ splits into product of linear factors, say $\delta=v_1\cdot v_2\cdot v_3$. Notice that

$$\mathrm{Sq}^1\delta=(v_1+v_2+v_3)\cdot\delta=0 \quad \text{implies} \quad v_1+v_2+v_3=0,$$

$$\text{i.e.,}\quad \delta=v_1v_2(v_1+v_2), \quad\text{and}\quad \mathrm{Sq}^2\delta=\gamma\cdot\delta \quad\text{implies}\quad \gamma=v_1^2+v_1v_2+v_2^2.$$

Again, it follows from Theorem B that $a(f_1)=\delta^{(k_1-1)}\cdot\prod_{j\neq 1}\alpha_j^{k_j}$ splits, and hence all $\alpha_j, j\neq 1$, split into product of linear factors, say

$$\alpha_j=w_j\cdot(w_j+\mu_{j,1})(w_j+\mu_{j,2})\cdot(w_j+\mu_{j,3}).$$

On the other hand,

$$\mathrm{Sq}^1\alpha_j=(\mu_{j,1}+\mu_{j,2}+\mu_{j,3})\cdot\alpha_j=0$$

implies that $\mu_{j1}+\mu_{j2}+\mu_{j3}=0$

$$\mathrm{Sq}^2\alpha_j=(\mu_{j1}\mu_{j2}+\mu_{j2}\mu_{j3}+\mu_{j3}\mu_{j1})\cdot\alpha_j=\gamma\cdot\alpha_j$$

implies that

$$\{\mu_{j1},\mu_{j2},\mu_{j3}\}=\{v_1,v_2,v_1+v_2\}$$

as a set, namely

$$\alpha_j=w_j(w_j+v_1)(w_j+v_2)(w_j+v_1+v_2).$$

(ii) *Suppose all* $k_j=1$: Then $s=(n+1)>1$, $a(f_1)=\prod_{j\neq 1}\alpha_j$. Hence, α_j again splits and the above proof will show that

$$\alpha_j=w_j(w_j+\mu_{j1})(w_j+\mu_{j2})(w_j+\mu_{j3}), \qquad \mu_{j1}+\mu_{j2}+\mu_{j3}=0,$$

and

$$(\mu_{j1}\mu_{j2}+\mu_{j2}\mu_{j3}+\mu_{j3}\mu_{j1})=\gamma \quad\text{for all}\quad j>1,$$

and the assertion follows easily. □

(C) *System of 2-Weights for Transformation Groups on $\mathbb{Z}_2$-Cohomology Projective Space*

Definition 1 (RP-case). In case the G-space $X\sim_2 RP^n$ and G is a $\mathbb{Z}_2$-torus, the *system of 2-weights* of X, $\Omega_2(X)$, is simply the set of roots with multiplicities of the structural equation $f(\xi)=0$ in Theorem (VI.7), namely,

$$\Omega_2(X)=\{w_j,k_j;\ 1\leqslant j\leqslant s\}.$$

Observe that the structural equation $f(\xi)=0$ depends on the choice of the lifting ξ, hence the above system of 2-weights is well-defined up to a translation in $H^1(B_G)$.

In general, suppose X is a $\mathbb{Z}_2$-CRP with a given action of a compact Lie group G. Let $\{H_i, 1\leqslant i\leqslant h\}$ be a complete set of representatives, one from each of the h conjugacy classes, of maximal $\mathbb{Z}_2$-tori of G. Then the system of 2-weights of X, $\Omega_2(X)$, consists of h subsystems, namely

$$\Omega_2(X)=\{\Omega_2(X\,|\,H_i);\ 1\leqslant i\leqslant h\}\,.$$

Definition 2 (CP-case). In case the G-space $X\sim_2 CP^n$ and G is a $\mathbb{Z}_2$-torus, there are two types of actions according to $\mathrm{Sq}^1(\eta)=\beta\cdot\eta=0$ or not. In the case $\beta=0$, $\Omega_2(X)=\{w_j, k_j;\ 1\leqslant j\leqslant s\}$ where w_j^2 are roots of $f(\eta)=0$. In the case $\beta\neq 0$, the roots of $f(\eta)=0$ is of the form $\beta w_j+w_j^2$. Observe that $\beta w_j+w_j^2=\beta(\beta+w_j)+(\beta+w_j)^2$, so that the w_j are only determined modulo β. Hence, one may write $G=\mathrm{Ker}(\beta)\times G/\mathrm{Ker}(\beta)=G^1\otimes\mathbb{Z}_2$ and assume that all the w_j lie in $H^1(B_{G'})\subseteq H^1(B_G)$. Then, we write the system of 2-weights of X in the following form

$$\Omega_2(X)=\beta\otimes\{w_j, k_j;\ 1\leqslant j\leqslant s\}$$

to suggest that the linear model of X is obtained from the tensoring of β and φ with $\Omega(\varphi)=\{w_j, k_j;\ 1\leqslant j\leqslant s\}$. (cf. Example 2b). Again, we shall define the *system of 2-weights* for a $\mathbb{Z}_2-CCP^n, X$, with a given action of compact Lie group G, to be the collection of h-subsystems, one for each conjugacy class of maximal $\mathbb{Z}_2$-tori, namely

$$\Omega_2(X)=\{\Omega_2(X\,|\,H_i)\}\,,$$

where $\{H_i, 1\leqslant i\leqslant h\}$ is a complete set of representatives of maximal $\mathbb{Z}_2$-tori.

Definition 3 (QP-case). In case the G-space $X\sim_2 QP^n$ and G is a $\mathbb{Z}_2$-torus, we define

$$\Omega_2(X)=\begin{cases}\{w_j, k_j;\ 1\leqslant j\leqslant s\} & \text{when } \mathrm{Sq}^2(\zeta)=0 \text{ and } \{w_j^4\} \text{ are roots of } f(\zeta)=0,\\ v\otimes\{w_j, k_j;\ 1\leqslant j\leqslant s\} & \text{when } \mathrm{Sq}^2(\zeta)=v^2\zeta\neq 0 \text{ and}\\ & \quad \{(v^2w_j^2+w_j^4)\} \text{ are roots of } f(\zeta)=0,\\ \{v_1, v_2\}\otimes\{w_j, k_j;\ 1\leqslant j\leqslant s\} & \text{when } \mathrm{Sq}^2(\zeta)=(v_1^2+v_1v_2+v_2^2)\cdot\zeta \text{ and}\\ & \quad w_j(w_j+v_1)\cdot(w_j+v_2)\cdot(w_j+v_1+v_2) \text{ are roots of } f(\zeta)=0.\end{cases}$$

In the general case that G is a compact Lie group,

$$\Omega_2(X)=\{\Omega_2(X\,|\,H_i);\ 1\leqslant i\leqslant h\}\,.$$

Remarks. (i) Observe that the system of 2-weights for CP-case and QP-case are again determined up to a translation in $H^1(B_G)$ if the action is of type a and up to a translation in $H^1(B_{G'})$, $G'=\mathrm{Ker}(\beta)$ (resp. $\mathrm{Ker}(v)$) if the action is of type b.

(ii) In case G is a compact Lie group, let $N(H_i)$ be the normalizor of H_i in G and $W(H_i)=N(H_i)/H_i$, then $W(H_i)$ acts naturally via conjugation on $H^1(B_{H_i})$

and it is not difficult to show that the i-th subsystem $\Omega_2(X \mid H_i)$ can be chosen to be *invariant* under the action $W(H_i)$.

(iii) Similar to the case of odd primes, it is again possible to define secondary 2-weight systems.

Concluding Remarks. In the above sections, structure theorems for torus (resp. $\mathbb{Z}_p$-torus) actions on cohomology projective spaces (with suitable coefficients) are proved. Formally, such structure theorems are formulated in terms of the splitting of equivariant cohomology algebras and the "linear structural constants" involved in the splitting equivariant cohomology algebras enable as to define the so-called geometric weight system (resp. p-weight system). Geometrically, the structure theorems prove that the cohomological aspect of orbit structure of a general torus (resp. p-torus) action on a cohomology projective Space X *is the same as that of a suitable linear model.* Hence, as far as transformations of *elementary abelian groups* on *cohomology projective spaces* in concerned, the structure theorems of the above sections exactly succeed in establishing the satisfactory cohomological linearization theorems that we were looking for. However, the situation of transformations of general compact Lie groups on cohomology projective spaces is much more complicated and deserves further penetrating investigation. As was already remarked in Chapter IV and also in Chapter V, similar general structure theorems of cohomological linerization type are *false in general.* However, it was demonstrated in Chapter V and in § 1 of this chapter by various testing problems that various *special* structure theorems (under suitable geometric simplicity conditions) are, in fact, provable by means of the maximal tori theorem and the structure theorems for torus actions. Roughly speaking, the maximal torus theorem implies that the orbit structure of the action of a compact connected Lie group G on X and the orbit structure of the restriction to a maximal torus T of G are closely related. Hence, the cohomological linerization theorem for torus actions will certainly imply that the orbit structure of the original G-actions also strongly resembles that of suitable linear models. This is the basic idea of the approach of geometric weight system, and there are many interesting problems that can be formulated and solved by this approach.

Chapter VII. Transformation Groups on Compact Homogeneous Spaces

The existence of abundant linear actions and the simplicity of topological structure form exactly the ideal combination that makes euclidean spaces, disks, spheres and projective spaces the best testing spaces for the study of transformation groups. So far, most of the deep results in topological transformation groups are still largely concentrating in the study of the above testing spaces. Generally speaking, the ideal combination of topological simplicity and abundant linear actions of the testing spaces are certainly very helpful in obtaining some basic understandings to *begin* with. For example, it is exactly the classical linear representation theory and those specific results of Chapters V and VI concerning such testing spaces that lead us to the basic understanding of the *central importance of elementary abelian groups* in the whole theory of topological transformation groups as well as to the formulation of those fundamental splitting theorems of Chapter IV (for actions of elementary abelian groups) in the setting of equivariant cohomology theory. However, the perspective will be undesirably limited if one becomes self-indulgent in insisting on the cohomological simplicity of the testing spaces. In this chapter, we begin to broaden the domain of testing spaces by considering transformation groups on compact homogeneous spaces. Since compact homogeneous spaces cover a wide range of topological types but still accommodate a rich variety of natural actions, they are particularly suitable for the study of transformation groups. However, strictly speaking, a systematic study of transformation groups on compact homogeneous spaces has not really gotten started yet. Hence, the few primitive results that we shall report in this chapter are still in a rather unsatisfactory form, which, in fact, should be merely considered as some preliminary indication of the abundance of interesting natural problems in this area.

To begin with, let us formulate some natural problems in the realm of transformation groups on compact homogeneous spaces as follows:

(A) *Transitive Actions and Classification of Compact Homogeneous Spaces in Terms of Their Diffeomorphic (resp. Topological, Homotopic or Rational Homotopic) Types*

Compact homogeneous spaces are by definition those manifolds with given *transitive differentiable* actions of compact Lie groups. Hence, it is natural to classify such transitive actions on a given compact homogeneous manifold M.

Suppose M is given as the coset space G/H. The given *transitive* G-action (via left translations) is said to be *primitive* if the restriction to any *proper normal* subgroup of G is no longer transitive. It is clear that the classification of transitive actions on a given compact homogeneous manifold M can be easily reduced to the classification of primitive transitive actions.

Problem 1. Suppose $M=G/H$ is a coset of a *simple* compact Lie group G. Classify primitive transitive actions on M. For most cases, one would expect that the original G-action is the *only primitive transitive* action on M.

Remarks. (i) Classification of transitive actions on spheres (homotopy spheres) was initiated by Montgomery and Samelson [M 3] and later completed by A. Borel [B 9]. The results may be summarized as follows: A homogeneous homotopy sphere must be standard and besides those well known transitive actions of classical groups on spheres, one has the following additional cases: $S^6=G_2/SU(3)$, $S^7=\mathrm{Spin}(7)/G_2$ and $S^{15}=\mathrm{Spin}(9)/\mathrm{Spin}(7)$.

(ii) Transitive actions on *integrál homology* spheres were classified by Bredon [B 17]. The Poincaré sphere, $SO(3)/A_5$, is the only homogeneous integral homology sphere which is not standard.

(iii) Transitive action on homogeneous spaces with non-vanishing Euler characteristic were essentially classified by H.C. Wang [W 2] and A. Borel and Siebenthal [B 12].

(iv) One may define the "rank" of a topological space X to be the sum of ranks of odd dimension homotopy group, $\sum \mathrm{rk}(\pi_{2k-1}(X))$. In [O 2] Oniscik classified transitive actions on homogeneous spaces of "rank" 1 and 2.

(v) Transitive actions on many Stiefel manifolds (real, complex or quaternionic) have been classified in [H 20]. The result is that of the expected uniqueness.

Classification of transitive (differentiable or topological) actions on compact homogeneous spaces essentially amounts to the classification of compact homogeneous spaces in terms of their diffeomorphic or topological types. In order to classify homogeneous spaces in terms of their topological or diffeomorphic types, it is technically useful and conceptually interesting to find *explicit* relationships between the infinitesimal data of the group pair (G,H) (or the Lie algebra pair $(\mathfrak{g},\mathfrak{h})$) and the topological or diffeomorphic invariants of G/H. Examples of such relationships are: (For simplicity, assume H connected.)

(i) $\mathrm{rk}(G)-\mathrm{rk}(H)=-\sum(-1)^k\,\mathrm{rk}(\pi_k(G/H))$.

(ii) It was proved in [O 2] that

$$P(G,t)/P(H,t)=\prod_{k=1}^{\infty}(1+t^{2k-1})^{(r_{2k-1}-r_{2k})}$$

where $P(G,t)$, $P(H,t)$ are the Poincaré polynomials of G and H respectively and $r_i=\mathrm{rk}(\pi_i(G/H))$.

(iii) According to Borel-Hirzebruch [B 11], the rational Pontrjgin classes of a compact homogeneous spaces G/H can be effectively computed in terms of the infinitesimal data of the pair (G,H); and it was recently proved by Novikov that rational Pontrjgin classes are topological invariants. It seems to the author that the rational Pontrjgin classes are going to be of increasing importance in the topological classification of homogeneous spaces.

Problem 2. Is it true that two homeomorphic compact homogeneous spaces are necessarily diffeomorphic? [In the extremely special case of spheres. the positive answer of above problem means that only the standard spheres have homogeneous differentiable structures.]

Remark. For most cases of classical homogeneous spaces such as Stiefel manifolds, Grassman manifolds, it is in fact not difficult to show that their homotopic types already uniquely characterize them among homogeneous spaces. The fascinating part of the above problem is how to give a systematic proof that will cover a wide range of homogeneous spaces.

Problem 3. In the study of compact transformation group, it is rather natural and useful to define the *torus rank of a space* X to be the *maximal dimension* of those tori that *act* almost freely on X. Is it true that the *torus rank of* G/H is equal to $(\mathrm{rk}(G)-\mathrm{rk}(H))$? [The special case $\mathrm{rk}(H)=0$ was proved by Allday in his thesis [A 3].]

(B) *Degree of Symmetry of Compact Homogeneous Spaces*

The following definition of degree of symmetry was first introduced in [H 11] to study the degree of symmetry of exotic spheres:

Definition. Let M be a smooth manifold and $\mathrm{Diff}(M)$ be the group of all diffeomorphisms of M onto itself. The *degree of (differentiable) symmetry* of M, denoted by $N_d(M)$, is defined to be

$$N_d(M)=\mathrm{Max}\,\{\dim G \text{ of all compact subgroups } G\subseteq \mathrm{Diff}(M)\}\,.$$

Similarly, one may also define the *degree of topological symmetry* of a space X, denoted by $N_t(X)$, to be

$$N_t(X)=\mathrm{Max}\,\{\dim G \text{ of all compact subgroups } G\subseteq \mathrm{Top}(X)\}$$

where $\mathrm{Top}(X)$ is the group of all homeomorphisms of X onto itself. Obviously $N_d(M)\leqslant N_t(M)$.

Examples. (i) For exotic spheres, $\sum^m$, it was proved in [H 11, H 6] that

$$N_d(\textstyle\sum^m)\leqslant\frac{1}{8}(m^2+7)<N_t(\sum^m)=N_t(S^m)=\frac{1}{2}(m^2+m)$$

and $N_d(\sum^m)=\frac{1}{8}(m^2+7)$ when and only when $\sum^m$ is a Kervaire sphere. In case $\sum^m\in bP_{4k}$, i.e., $m=(4k-1)$ and $\sum^m$ bounds a parallelizable manifold, then it is known that $N_d(\sum^m)=\frac{1}{8}(m^2-4m+11)$. We refer the reader to [H 9, H 18] for various interesting improvements of estimates of $N_d(\sum^m)$ for those exotic spheres which do not bound parallelizable manifolds, i.e., $\sum^m\notin bP_{(m+1)}$.

(ii) Let X be a *smoothable topological* manifold. Then it is obvious that

$$N_t(X)\geqslant\mathrm{Max}\,\{N_d(M)\colon M \text{ are smoothings of } X\}$$

and equality holds for most topological manifolds *with smoothings of sufficient high degrees of symmetry*. However, the following example suggests that there should exist plenty of examples with $N_t(X) \gneq \mathrm{Max}\{N_d(M); |M|=X\}$, when $|M|$ denotes the underlying topological manifold of M:

Let E_1, E_2 be two copies of the tangent disc bundle of S^{2k+1}. The orthogonal action of $0(2k+1)$ on S^{2k+1} induces natural actions on E_1, E_2 respectively. Choose an *invariant* D^{2k+1} around one of the two fixed points, say $p \in S^{2k+1}$. Then $\pi^{-1}(D^{2k+1}) \subseteq E_i$ is clearly equivariantly diffeomorphic to $D^{2k+1} \times D^{2k+1}$ with the usual diagonal $0(2k+1)$-action: $g \cdot (x,y) = (g \cdot x, g \cdot y)$. Hence the following identification map, $i(x,y)=(y,x)$, is obviously *equivariant*, and therefore, the *plumbing of* E_1, E_2 which is a $(4k+2)$-dimensional manifold M obtained by identifying the $D^{2k+1} \times D^{2k+1}$ part of E_1 with the $D^{2k+1} \times D^{2k+1}$ part of E_2 via $i(x,y)=(y,x)$. Let $C(\partial M)$ be the cone over ∂M and

$$X = M \cup_f C(\partial M)$$

where f identifies the base of $C(\partial M)$ with $\partial M \subseteq M$. Then X is a *topological* manifold with an $0(2k+1)$-action fixing the vertex of the cone $C(\partial M)$. For most values of k, it is well known that X is a *non-smoothable topological* manifold. On the other hand, it is not too difficult to prove that $N_t(X) = \dim 0(2k+1) = k(2k+1)$.

(iii) Suppose the first rational Pontrjgin class of M vanishes, i.e., $P_1(M)=0$. Then the result of [H 5] becomes a useful tool in the study of the degree of symmetry of such manifolds. For example, it was proved in [H 18] that $N_d(M^{m_1} \times M^{m_2} \times \cdots \times M^{m_k}) \leqslant N_d(S^{m_1} \times S^{m_2} \times \cdots \times S^{m_k}) = \sum_{i=1}^k N_d(S^{m_i}) = \frac{1}{2}\sum (m_i^2 + m_i)$, under the assumption that $P_1(M^{m_1} \times \cdots \times M^{m_k})=0$. Notice that the rational Pontrjgin classes are topological invariants. Hence the above condition $P_1(M^{m_1} \times \cdots \times M^{m_k})=0$ is a topological condition and it is natural to pose the following problem:

Problem 4. Is it true that

$$N_t(M^{m_1} \times \cdots \times M^{m_k}) \leqslant N_t(S^{m_1} \times \cdots \times S^{m_k}) = \sum_{i=1}^k N_t(S^{m_i}) = \tfrac{1}{2}\sum_{i=1}^k (m_i^2 + m_i)$$

under the assumption $P_1(M^{m_1} \times \cdots \times M^{m_k})=0$?

Problem 5. Suppose $M = G/H$ is a compact homogeneous space and G is one of the *highest dimensional, effective, transitive*, transformation groups on M. Is it true that $N_t(M) = N_d(M) = \dim G$?

There are the following variants of the concept of degree of symmetry which are closely related to the original one but are technically simpler to handle.

Definition. The *differentiable torus-degree of symmetry* of M is defined to be $T_d(M) = \mathrm{Max}\{\dim T;$ all subtori $T \subseteq \mathrm{Diff}(M)\}$. And similarly, the *topological torus-degree of symmetry* of X is defined to be

$$T_t(M) = \mathrm{Max}\{\dim T; \text{ all subtori } T \subseteq \mathrm{Top}(X)\}.$$

Definition. Consider the actions of *simple* compact Lie groups instead of *general* compact Lie groups, it is natural to define

$$\mathrm{SN}_d(M) = \mathrm{Max}\{\dim G;\ G \text{ simple compact Lie subgroups of } \mathrm{Diff}(M)\},$$
$$\mathrm{SN}_t(M) = \mathrm{Max}\{\dim G;\ G \text{ simple compact Lie subgroups of } \mathrm{Top}(X)\}.$$

Problem 5'. Suppose $M = G/H$ and G is an *effective, transitive* transformation group on M of *highest rank*. Is it true that $T_d(M) = T_t(M) = \mathrm{rk}(G)$?

Problem 5''. Suppose $M = G/H$ and G is an *effective, transitive* transformation group on M of *highest dimension*; G_1 is one of the *highest dimensional simple normal* factor of G. Is it true that $\mathrm{SN}_d(M) = \mathrm{SN}_t(M) = \dim G_1$?

Remark. Problem 5' and 5'' are technically more accessible than problem 5, and they are, in fact, stepping stones toward the eventual solution of problem 5.

(C) *Actions of "Large" Compact Lie Groups on Homogeneous Spaces*

Roughly speaking, an effective transformation group K on a compact homogeneous spaces $M = G/H$ is considered to be a "large" transformation group on M if $\dim(K)$ is not too small as compared to $\dim(G)$. For examples $\mathrm{SO}(k)$ with $k > \frac{1}{2}n$ are considered to be large transformation groups on Stiefel manifolds $V_{n,l} = O(n)/O(n-l)$.

In the case of irreducible compact symmetric spaces $M = G/K$, the orbit structure of the isometric transformation of the isotropy group K on M is known to be quite simple and useful ever since É. Cartan. For example, take the case of the group space $M = G = (G \times G)/G$ of a *simple compact connected Lie group G*. Then the isometric transformations of the *isotropy* group G on the group space G itself are just the conjugation transformations and the orbit structure of such an action is the important geometric decomposition of G into conjugacy classes given in § 1 of Chapter II.

Problem 6. Let $M = G/K$ be an irreducible compact symmetric space. Is it true that all the *differentiable actions* of K on M must have the same cohomological orbit structure as that of the *isometric* isotropy action? Of course, in the special case that $M = G = (G \times G)/G$ is the group manifold of a simple compact connected Lie group G, one should assume that the differentiable G-action is *non-transitive*.

Problem 7. Suppose $M = G/H$ and $G = \mathrm{SO}(n)$ (resp. $\mathrm{SU}(n)$, or $\mathrm{Sp}(n)$) is a classical group. Let $K = \mathrm{SO}(k)$ (resp. $\mathrm{SU}(k)$, or $\mathrm{Sp}(k)$), $k > \frac{1}{2}n$, and there are *no subgroups* of $Z(H, G)$ *isomorphic to* K. It is true that the orbit structure of any K-action on M must be largely the same as that of the left translations restricted to $K \subseteq G$?

In this chapter, we shall apply the cohomological method developed in the previous chapters as the main tool to investigate a few relatively simple cases of the above problems. The results are rather primitive and far from being satisfactory. However, they seems to indicate the existence of plenty of natural, deep problems and to offer a preliminary justification of taking compact homogeneous spaces as the main family of testing spaces for the study of transformation groups.

Moreover, they also seem to confirm the general feeling that the cohomology theory of transformation groups (especially those splitting theorems of Chapter IV) is so far one of the most powerful tools in the study of orbit structures of transformation groups on spaces of a given type.

§ 1. Topological Transformation Groups on Spaces of the Rational Homotopy Type of Product of Odd Spheres

In this section, we shall study actions of compact connected Lie groups, G, on cohomology manifolds, X, of the rational homotopy type of product of odd spheres. There are many interesting examples of such G-spaces besides the usual examples of products of odd spheres with "linear" G-actions. For examples, the spaces of compact connected Lie groups themselves, the Stiefel manifolds, $SU(2n)/Sp(n)$, etc., are all manifolds of the rational homotopy type of product of odd spheres and are capable of accommodating various interesting transformation groups. Since the cohomological machinary developed in the previous chapters is going to be the main tool of the following investigation, let us begin with simple computations of the possibilities of equivariant cohomology algebras:

(A) *Computations of* $H_G^*(X,k)$

Let G be a compact connected Lie group and X be a G-manifold of the rational homotopy type of a product of odd spheres. Let $\{x_1, \ldots, x_l\}$ be (homogeneous) basis of $\pi_*(X) \otimes k$ and $\{y_1, \ldots, y_r\}$ be (homogeneous) basis of $\pi_*(B_G) \otimes k$, $\dim x_i = m_i$ are odd and $\dim y_j = n_j$ are even, $r = \mathrm{rk}(G)$. Then, it is well-known that $R_G = H^*(B_G, k) \cong k[y_1, \ldots, y_r]$ and there is an associated graded differential algebra of X_G, say $\{L(X_G), d\}$, satisfying the following properties:

(i) $L(X_G) \cong \Lambda[x_1, \ldots, x_l] \otimes_k R_G$ (as a graded algebra),

(ii) d is an anti-derivation of degree 1, $d|R_G = 0$, $d^2 x_i = 0$ and the restriction of dx_i to $\Lambda[x] \otimes_k R_G^0 = \Lambda[x]$ are zero for $1 \leq i \leq l$,

(iii) $H(L(X_G), d) \cong H_G^*(X)$ as an R_G-algebra,

(iv) for any connected subgroup $K \subseteq G$, the induced homomorphism

$$\iota^*: L(X_G) = \Lambda[x] \otimes R_G \to \Lambda[x] \otimes R_K = L(X_K)$$

commutes with the differentials and induces the usual homomorphism: $H_G^*(X) \to H_K^*(X)$ at the homology level.

The above graded differential algebra $\{L(X_G), d\}$ is one of the basic machinary for the computation of $H_G^*(X)$. The "coefficients" in the expression of dx_i in terms of lower dimensional exterior products of x's are elements of R_G^+ which depend on the homotopy type of X_G.

Example 1.1. In the very special case of $X = G$ with G acting transitively as left translations, $X_G = E_G$ and

$$L(X_G) = \Lambda[x_1, \ldots, x_r] \otimes k[y_1, \ldots, y_r]$$

with $dx_i = y_i$. This is exactly the well-known transgression theorem of A. Borel [B 2]. Now, suppose K is a connected subgroup of G. Then there is a fibration $E_K \to X_K \to X/K = G/K$ with contractible fibre E_K. Hence X_K is of the same homotopy type as G/K and its associated graded differential algebra:

$$L(X_K) = \Lambda[x_1, \dots, x_r] \otimes R_K; \qquad d'x_i = \iota^*(dx_i) = \iota^*(y_i) \in R_K$$

is exactly the well-known Koszul complex with $H(L(X_K), d) = H^*(X_K) = H^*(G/K)$.

Example 1.2. Suppose $X = X_{n,l} = \mathrm{Sp}(n)/\mathrm{Sp}(n-l)$ is a quaternionic Stiefel manifold and $G = \mathrm{SP}(n)$ acts on X via left translation. Then $X_G = B_{\mathrm{SP}(n-l)}$ and the associated graded differential algebra is given as follows:

$$L(X_G) = \Lambda[x_1, \dots, x_l] \otimes k[y_1, \dots, y_n], \qquad dx_i = y_i$$

where $\dim x_i = 4(n-i)+3$, $\dim y_j = 4(n-j+1)$. Again, if we restrict the G-action on $X_{n,l}$ to a connected subgroup $K \subseteq G$, then we have

$$L(X_K) = \Lambda[x_1, \dots, x_l] \otimes R_K \quad \text{and} \quad d'x_i = \iota^*(y_i) \in R_K .$$

For example, if $K = \mathrm{Sp}(k)$, $k > \frac{1}{2}n$, then $\iota^*(y_i) = 0$ for $1 \leqslant i \leqslant (n-k)$ and $\iota^*(y_i) = y'_{(k-n+i)} \in R_K = k[y'_1, \dots, y'_k]$.

Example 1.3. Suppose X is of the rational homotopy type of a product of *two odd spheres*. Then $L(X_G) = \Lambda[x_1, x_2] \otimes R_G$ and $dx_i = a_i \in R_G$. Let b be the g.c.d. of a_1, a_2; $b \cdot a'_1 = a_1$, $b \cdot a'_2 = a_2$. Then it is easy to compute that $H^*_G(X) = H(L(X_G), d) \cong R_G[\xi]/\langle a_1, a_2, b\xi \rangle$ where ξ is represented by the "cycle" $(a'_1 x_2 - a'_2 x_1)$. Hence the annahilator ideals of 1 and ξ in $H^*_G(X)$ are respectively $\langle a_1, a_2 \rangle$ and $\langle b \rangle$. Let $T \subseteq G$ be a maximal torus of G. Since the induced homomorphism $\iota^*: R_G \to R_T$ is injective, it is clear that $H^*_T(X) \cong R_T[\xi]/\langle a_1, a_2, b\xi \rangle$. Therefore, it follows from Theorem IV.8 that b splits into a product of linear factors in R_T and the prime ideals in R_T belonging to $J = \mathrm{Ann}(1) = \langle a_1, a_2 \rangle$ must be linear.

Remarks. (i) Generally speaking, the above graded differential algebra $\{L(X_G), d\}$ is a convenient machinery to set up the beginning step of the computation of $H^*_G(X)$. However, the structural invariants of $H^*_G(X)$ (such as $\langle b \rangle$ and $\langle a_1, a_2 \rangle$ in the above example (1.3)) that one gets from $\{L(X_G), d\}$ are usually too general. An important refinement which imposes strong restrictions on those structural invariants can be obtained by applying those splitting theorems of Chapter IV to $H^*_T(X)$. [Notice that $H^*_G(X) = H^*_T(X)^W$, and $H^*_T(X) = H^*_G(X) \otimes_{R_G} R_T$ (cf. Prop. 1 of § 1, Ch. III)].

(ii) In the study of transformation groups on a given space X via cohomology theory, the determination of the *possibilities* of $H^*_G(X)$ is actually the heart of the game, which usually involves several steps of successive *refinements* of those *structural constants* of $\{L(X_G), d\}$.

(iii) In the case that X is of the rational homotopy type of product of odd spheres, the associated graded differential algebra of X_G, $\{L(X_G), d\}$, is simply an exterior algebra over R_G with a differential. Therefore, the computation of $H^*_G(X)$

$=H(L(X_G),d)$ looks like a rather straightforward simple algebraic problem. However, so far, there is very little general understanding of this problem. The following are some simple general facts concerning $H(L(X_G)d)$:

Lemma 1.1. *Let* $L(X_G)=\Lambda[x_1,\ldots,x_l]\otimes R_G$ *be an exterior algebra over* R_G *with a given differential. Let* $a_i\in R_G$ *be the constant term of* dx_i, i.e., $dx_i=a_i+$ *terms involving* $x's\,(1\leqslant i\leqslant l)$. *Then, the annahilating ideal of* $1\in H(L(X_G),d)$ *and the ideal generated by* $\{a_1,\ldots,a_l\}$ *have the same radical in* R_G.

Proof. Let $J=\mathrm{Ann}(1)$ be the annahilator of $1\in H(L(X_G),d)$. We shall show that

$$\langle a_1^l,\ldots,a_l^l\rangle\subseteq J\subseteq\langle a_1,\ldots,a_l\rangle$$

which clearly implies that $\sqrt{J}=\sqrt{\langle a_1,\ldots,a_l\rangle}$. The second inclusion $J\subseteq\langle a_1,\ldots,a_l\rangle$ is rather obvious, let us prove that first inclusion as follows: Let

$$dx_i=a_i+g_i(x)\,,\qquad a_i\in R_G\,,\qquad 1\leqslant i\leqslant l$$

where $g_i(x)$ are of positive "exterior degree" and hence $[g_i(x)]^l=0$. Then $d(g_i(x))=d^2x_i-da_i=0$ implies $d(a_i^{l-1}-a_i^{l-2}\cdot g_i(x)+\cdots+[-g_i(x)]^{l-1})=0$. Therefore

$$d\{x_i\cdot[a_i^{l-1}-a_i^{l-2}\cdot g_i(x)+\cdots+(-g_i(x))^{l-1}]=a_i^l-(g_i(x))^l=a_i^l\in J\,.\quad\square$$

Remark. It follows from the above lemma that J and $\langle a_1,\ldots,a_l\rangle$ have the same set of minimal prime ideals (in R_T) which are proved in Theorem (IV.6) to be exactly those linear ideals corresponding to the *maximal* connected isotropy subgroups of the restricted action of the maximal torus T of G. Hence, as a corollary of Theorem (IV.6) and the above Lemma (1.1), we have reproduced the following result of Allday [A 3]:

Theorem (VII.1). *Let X be a G-space of the rational homotopy type of a product of l odd spheres. Then*

$$\mathrm{Max}\{\mathrm{rk}(G_x);\,x\in X\}\geqslant(\mathrm{rk}(G)-l)\,.$$

Proof. Since $\langle a_1,\ldots,a_l\rangle$ is generated by l elements, the codimensions of its varieties (which are automatically linear) are at most l. Hence the above theorem follows directly from Theorem (IV.6). $\square$

Lemma (1.2). *Suppose* $X\sim_k S^{m_1}\times\cdots\times S^{m_l}$ *is of the rational homotopy type of product of odd spheres and moreover*

$$\mathrm{Max}\{(m_i+1);\,1\leqslant i\leqslant l\}\leqslant\mathrm{Min}\{(m_i+m_j),\,1\leqslant i\leqslant j\leqslant l\}\,.$$

Then $L(X_G)=\Lambda[x_1,\ldots,x_l]\otimes R_G$ *and* $dx_i=a_i\in R_G$.

Proof. It follows immediately from the above dimensional restriction that any elements of dimension (m_i+1), such as dx_i, must be in R_G. Hence $dx_i=a_i\in R_G$. $\square$

(B) *Connectedness of* $F(T, X)$ *and its Consequences*

For spaces with vanishing even-dimensional rational homotopy groups, i.e., $\pi_{2j}(X)\otimes k=0$, there is the following useful theorem which asserts the connectedness of $F(T,X)$:

Theorem (IV.5). *Let T be a torus group and X be a T-space with vanishing even-dimensional rational homotopy groups. Then the fixed point set F (if non-empty) is always connected.*

In this subsection, we shall deduce several interesting consequences of the above theorem:

Theorem (VII.2). *Let* $G=\mathrm{SO}(n)$ *(resp.* $\mathrm{SU}(n)$, $\mathrm{Sp}(n)$*) and* X *be a cohomology G-manifold with vanishing even-dimensional rational homotopy groups. Suppose that* $F(T,X)\neq\emptyset$ *and the system of "local weights" around* $F(T,X)$ *is given by* $\Omega_F(X)=\{\pm\theta_i; l\}$. *Then all connected isotropy subgroups* G_x^0, *are of the form* $\mathrm{SO}(n_x)$ *(resp.* $\mathrm{SU}(n_x)$, $\mathrm{Sp}(n_x)$*) with* $n\geqslant n_x\geqslant(n-l)$.

Proof. The proof of the above theorem is essentially the same as that of Theorem (V.2) [cf. § 2-B, Ch. V]. Let T be a maximal torus of G and $(\theta_1,\ldots,\theta_r)$ be the usual coordinates of T. It follows from Theorem (IV.5) that $F(T,X)$ is connected and hence the system of local weights $\Omega_F(X)$ is uniquely defined. Moreover, the fixed point sets of subtori $S\subseteq T$, $F(S,X)$, are also connected. Let $x\in X$ be an arbitrary point and $F(x)=F(T_x^0,X)$ be the F^0-variety of x in X. Up to a conjugation, we may assume $(G_x\cap T)^0=T_x^0$ is a maximal torus of G_x^0. Since $F(x)\supseteq F(T,X)$, it follows from Theorem (V.1′) that

$$T_x^0=(G_x\cap T)^0=\theta_{j1}^{\perp}\cap\cdots\cap\theta_{jt}^{\perp},$$

and up to a conjugation by element of Weyl group $W(G)$, we may assume that $T_x^0=\theta_1^{\perp}\cap\theta_2^{\perp}\cap\cdots\cap\theta_t^{\perp}$. Now, it follows from Corollary 2 of Proposition 2 of § 2-A, Ch. V that

$$\Delta(G_x^0)\supseteq\Delta(G)\,|\,T_x^0\setminus\Omega_F(X)|\,T_x^0,$$

and it is an easy exercise of Lie algebra to show that

$$G_x^0=\mathrm{SO}(n_x),\quad \left[\frac{n_x}{2}\right]=(r-t)\qquad(\text{resp. }\mathrm{SU}(n-t),\ \mathrm{Sp}(n-t)).\quad\square$$

Corollary 1. *Let* $G=\mathrm{SO}(n)$ *(resp.* $\mathrm{SU}(n)$, $\mathrm{Sp}(n)$*),* $n\geqslant 10$, *and X be a topological G-manifold with vanishing first rational Pontrjgin class and vanishing even-dimensional rational homotopy groups. Suppose*

$$\dim H_X>\tfrac{1}{3}\dim G$$

where (H_X) *is the principal isotropy subgroup type of X. Then, again, all connected isotropy subgroups* G_x^0 *are of the type* $\mathrm{SO}(n_x)$ *(resp.* $\mathrm{SU}(n_x)$, $\mathrm{Sp}(n_x)$*).*

Proof. It follows from the slice theorem that a principal orbit $G(x_0)=G/H_X \subseteq X$ imbeds with trivial normal micro bundle. Hence, it follows from $P_1(X,k)=0$ that $P_1(G/H_X;k)=0$. Then, it is not difficult to improve the proof of [H 5] to show that $P_1(G/H_X;k)=0$ and $\dim H_X > \frac{1}{3}\dim G$ imply that $H_X^0=\mathrm{SO}(l)$ resp. $\mathrm{SU}(l)$, $\mathrm{Sp}(l)$) with $l>\frac{3}{5}\cdot n$. Then, applying Theorem (VII.2) to the restricted action of $\mathrm{SO}(l)$ (resp. $\mathrm{SU}(l)$, $\mathrm{Sp}(l)$) on X, we have $G_x^0 \cap \mathrm{SO}(l)=\mathrm{SO}(l_x)$ [resp. $G_x^0 \cap \mathrm{SU}(l) =\mathrm{SU}(l_x)$, $G_x^0 \cap \mathrm{Sp}(l)=\mathrm{Sp}(l_x)$]. Since $l>\frac{3}{5}n$, it is again an easy exercise of Lie algebra to show that the only subgroups of G with the above intersection property are those subgroups of the form $\mathrm{SO}(n_x)$ (resp. $\mathrm{SU}(n_x)$, $\mathrm{Sp}(n_x)$). □

As an application of the above corollary, we have the following *topological version* of Theorem 2 of [H 18] (which was proved in differentiable setting):

Theorem (VII.3). *Let $X \sim_k S^{m_1}\times\cdots\times S^{m_l}$ be a closed topological manifold of the rational homotopy type of the product of odd spheres with dimension $m_j \geq 5$. Then the topological degree of symmetry of X.*

$$N_t(X) \leq \sum \tfrac{1}{2} m_j(m_j+1)$$

and equality holds only when $X=G/H$, $G=\mathrm{SO}(m_1+1)\times\cdots\times\mathrm{SO}(m_l+1)$ and $H^0=\mathrm{SO}(m_1)\times\cdots\times\mathrm{SO}(m_l)$.

Proof. In the proof of Theorem 2 of [H 18] for the *differentiable* case, the only part which *relies* on the *differentiability* of the actions is Lemma 2 [p. 11 of H 18]. However, the above Corollary 1 exactly proves the same assertion of Lemma 2 of [H 18] for the *topological* case. Hence, one simply replaces Lemma 2 of [H 18] by the above Corollary 1, then, the rest of the proof of the above theorem is the same as that of Theorem 2 of [H 18]. We refer the reader to [p. 12—15, H 18] for such a proof.

Next, let us consider the more restrictive case that $X \sim_{\mathbb{Z}} S^{m_1}\times\cdots\times S^{m_k}$ is of *integral cohomology type* of product of odd spheres satisfying the following dimensional restriction:

$$\mathrm{Max}\{(m_j+1);\, 1\leq j\leq k\} \leq \mathrm{Min}\{(m_i+m_j);\, 1\leq i\leq j\leq k\}.$$

Under this more restrictive conditions, one can imporve the conclusion of Theorem (VII.2) up to the isotropy subgroups themselves rather than the *connected* isotropy subgroups. Namely,

Theorem (VII.2′). *Let $G=\mathrm{SO}(n)$ (resp. $\mathrm{SU}(n)$, $\mathrm{Sp}(n)$) and $X\sim_{\mathbb{Z}} S^{m_1}\times\cdots\times S^{m_k}$; $\mathrm{Max}\{(m_j+1)\} \leq \mathrm{Min}\{(m_i+m_j)\}$. Suppose that G acts differentiably on X and $F(T,X)\neq\emptyset$; T is a maximal torus of G. If the system of local weights is given by $\Omega_F(X)=\{\pm\theta_i, l\}$ then all isotropy subgroups, G_x are of the form $\mathrm{SO}(n_x)$ (resp. $\mathrm{SU}(n_x)$, $\mathrm{Sp}(n_x)$).*

Proof. It follows from the above dimensional restriction and lemma (1.2) that $L(X_G)=\Lambda[x_1,\ldots,x_k]\otimes R_G$ and $dx_i=a_i\in R_G$. Since $F(T,X)\neq\emptyset$, $\pi^*: H^*(B_G) \to H^*(X_G)$ is injective and hence $dx_i=a_i=0$ for all $1\leq i\leq k$, implying $d\equiv 0$ and $H_G^*(X)=L(X_G)$. Then, it follows easily from Theorem (IV.1) that $F(T,X)$ is also of the *integral cohomology type of product of k odd spheres.* It has already been shown in Theorem (VII.2) that $F(T,X)=F(G,X)$ (or $F(T,X)=F(\mathrm{SO}(n-1),X)$ if $G=\mathrm{SO}(n)$ and n is odd) and G_x^0 are all of the form $\mathrm{SO}(n_x)$ (resp. $\mathrm{SU}(n_x)$, $\mathrm{Sp}(n_x)$). We shall make use of the additional assumption of *differentiability* and $X\sim_{\mathbb{Z}} S^{m_1}\times\cdots\times S^{m_k}$ to prove that all isotropy subgroups, G_x, are in fact *connected.* It follows from the *differentiability* and the assumption $\Omega_F(X)=\{\pm\theta_i,l\}$ that the G_x are connected for those x in a small neighborhood of $F(T,X)$, say $N(F(T,X))$. Now, suppose the contrary that there is at least one point $x_1\in X$ with *disconnected* isotropy subgroup $G_{x_1}\neq G_{x_1}^0$. Let K be a subgroup of G_{x_1} such that

$$G_{x_1}\supseteq K\supseteq G_{x_i}^0 \quad\text{and}\quad K/G_{x_1}^0\cong\mathbb{Z}_p,\quad p \text{ prime}.$$

By suitably restricting to $\mathrm{SO}(n')$ (resp. $\mathrm{SU}(n')$, $\mathrm{Sp}(n')$) for smaller n' if necessary, one may assume, without loss of generality, that there are *no proper* subgroups of the type $\mathrm{SO}(k)$ (resp. $\mathrm{SU}(k)$, $\mathrm{Sp}(k)$) *containing* K. Let K_p be the maximal p-torus of K. Then it is not difficult to see that

$$F(K_p,N(F(T,X)))=F(T,X).$$

Hence $F(K_p,X)\supseteq F(T,X)+\{x_1\}$ must be *disconnected.* Therefore,

$$\begin{aligned}\dim_{\mathbb{Z}_p}H^*(F(K_p,X);\mathbb{Z}_p)&>\dim_{\mathbb{Z}_p}H^*(F(T,X);\mathbb{Z}_p)\\&=\dim_{\mathbb{Z}_p}H^*(X;\mathbb{Z}_p)=2^k\end{aligned}$$

which is a contradiction to a theorem of A. Borel [B 10]. Hence, all isotropy subgroups G_x must be *connected.* □

Remark. The above proof is essentially the same as those of Theorems (3.1), (3.2), and (3.3) of [p. 745, H 5]. It seems to the author that the above theorem should still hold without the differentiability assumptions.

Corollary 1′. *Let* $G=\mathrm{SO}(n)$ (*resp.* $\mathrm{SU}(n)$, $\mathrm{Sp}(n)$), $n\geqslant 10$, *and* X *be a differentiable G-manifold of the integral cohomology type of product of odd spheres, and let* $P_1(X,k)=0$. *Suppose*

$$\dim H_X>\tfrac{1}{3}\dim G$$

where (H_X) *is the principal isotropy subgroup type of* X. *Then all isotropy subgroups,* G_x, *are connected and of the form* $\mathrm{SO}(n_x)$ (*resp.* $\mathrm{SU}(n_x)$, $\mathrm{Sp}(n_x)$).

(C) *Actions of Large Compact Lie Groups on Stiefel Manifolds*

Topologically, Stiefel manifolds are spaces of the same *rational homotopy type* of product of spheres but are actually non-homotopic to product of spheres (except for the very special cases such as $V_{8,2}=\mathrm{SO}(8)/\mathrm{SO}(6)=S^7\times S^6$). For most cases, the *difference* between Stiefel manifolds and product of spheres can be detected by the Steenrod square operations. Therefore, it is interesting to investigate how the differences with respect to Steenrod square operations will be reflected in the geometric behaviors of transformations groups on these manifolds. This is a natural testing problem in transformation groups where cohomology operations will surely play an essential role. So far, results in this direction are still rather unsatisfactory and are mostly as yet unpublished. However, in order to give a partial indication of the general type of results in this direction, we shall include here some of the simpler theorems as examples.

To begin with, let us recall the following well-known fact about square operations on mod 2 characteristic classes and mod 2 cohomologies of Stiefel manifolds:

Let $\mathbb{Z}_2^n \subseteq 0(n)$ be the $\mathbb{Z}_2$-maximal torus of $0(n)$. Then

$$H^*(B_{0(n)};\mathbb{Z}_2)=\mathbb{Z}_2[w_1,w_2,\ldots,w_n]\subseteq\mathbb{Z}_2[t_1,\ldots,t_n]=H^*(B_{\mathbb{Z}_2^n};\mathbb{Z}_2)$$

where w_j is the j-th universal Stiefel-Whitney class and is identified with the j-th basic symmetric polynomial of $(t_1,\ldots,t_n)$, $\deg t_i=1$. Similarly, one has

$$H^*(B_{U(n)};\mathbb{Z}_2)=\mathbb{Z}_2[c_1,c_2,\ldots,c_n]\subseteq\mathbb{Z}_2[t_1,\ldots,t_n]=H^*(B_{\mathbb{Z}_2^n};\mathbb{Z}_2)$$

where c_j is the mod 2 j-th universal Chern class and is identified with the j-th basic symmetric polynomial of $(t_1^2,t_2^2,\ldots,t_n^2)$; and

$$H^*(B_{\mathrm{Sp}(n)};\mathbb{Z}_2)=\mathbb{Z}_2[q_1,q_2,\ldots,q_n]\subseteq\mathbb{Z}_2[t_1,t_2,\ldots,t_n]=H^*(B_{\mathbb{Z}_2^n};\mathbb{Z}_2)$$

where q_j is the mod 2 j-th universal quaternionic class and is identified with the j-th basic symmetric polynomial of $(t_1^4,t_2^4,\ldots,t_n^4)$.

Proposition (1.1). *The square operations in* $H^*(B_{0(n)};\mathbb{Z}_2)$, $H^*(_{U(n)};\mathbb{Z}_2)$, $H^*(B_{\mathrm{Sp}(n)};\mathbb{Z}_2)$, $H^*(B_{\mathrm{SO}(n)};\mathbb{Z}_2)$, $H^*(B_{\mathrm{SU}(n)};\mathbb{Z}_2)$ *are given as follows:*

(i) $$\mathrm{Sq}^i w_j=\sum_{a=0}^{i}\binom{j-i+a-1}{a} w_{i-a}w_{j+a} \quad (i\leqslant j),$$

(ii) $\mathrm{Sq}^{2i+1}c_j=0$ *for all i,j and* $$\mathrm{Sq}^{2i}c_j=\sum_{a=0}^{i}\binom{j-i+a-1}{a} c_{i-a}c_{j+a} \quad (i\leqslant j),$$

(iii) $\mathrm{Sq}^k q_j=0$ *if* $k\not\equiv 0 \bmod 4$ *and* $$\mathrm{Sq}^{4i}q_j=\sum_{a=0}^{i}\binom{j-i+a-1}{a} q_{i-a}q_{j+a} \quad (i\leqslant j).$$

Moreover, (i) *and* (ii) *also hold for* $H^*(B_{\mathrm{SO}(n)};\mathbb{Z}_2)$ *and* $H^*(B_{\mathrm{SU}(n)};\mathbb{Z}_2)$ *respectively provided that one puts* $w_1=0$ *and* $c_1=0$.

Proof. Since $w_j=\sum t_1t_2,\ldots,t_j$ is the j-th basic symmetric polynomial of $(t_1,\ldots,t_n)$, it is clear that

$$\mathrm{Sq}^i w_j=\sum t_1^2\ldots t_i^2\, t_{i+1}\ldots t_j \quad (i\leqslant j)$$

is the symmetric polynomial generated by $t_1^2\ldots t_i^2\, t_{i+1}\ldots t_j$. Then it is a straightforward verification by induction on n that

$$\mathrm{Sq}^i w_j=\sum_{a=0}^{i}\binom{j-i+a-1}{a} w_{i-a}w_{j+a}.$$

The same formula with t_j replaced by t_j^2 and t_j^4 respectively also proves (ii) and (iii) for mod 2 Chern classes and mod 2 quaternionic classes. In the cases of $H^*(B_{SO(n)};\mathbb{Z}_2)$ and $H^*(B_{SU(n)};\mathbb{Z}_2)$, one simply observes that

$$H^*(B_{O(n)};\mathbb{Z}_2)\to H^*(B_{SO(n)};\mathbb{Z}_2)$$

maps w_1 to 0 and w_j to w_j for $j>1$;

$$H^*(B_{U(n)};\mathbb{Z}_2)\to H^*(B_{SU(n)};\mathbb{Z}_2)$$

maps c_1 to 0 and c_j to c_j for $j>1$.

Proposition (1.2). *Let* $V_{n,k}=0(n)/0(k)$, $W_{n,k}=U(n)/U(k)$ *and* $X_{n,k}=\mathrm{Sp}(n)/\mathrm{Sp}(k)$ *be respectively real, complex and quaternionic Stiefel manifolds. Then*

(i) $H^*(W_{n,k};\mathbb{Z}_2)\cong\Lambda_{\mathbb{Z}_2}[x_k,x_{k+1},\ldots,x_{n-1}]$, $\deg x_j=2j+1$ *and* $\mathrm{Sq}^{2i}x_j=\binom{j}{i}x_{j+i}$, $i\leqslant j,\ j+i\leqslant(n-1)$; *and* 0 *otherwise,*

(ii) $H^*(X_{n,k};\mathbb{Z}_2)\cong\Lambda_{\mathbb{Z}_2}[x_k,x_{k+1},\ldots,x_{n-1}]$, $\deg x_j=4j+3$, $\mathrm{Sq}^{4i}x_j=\binom{j}{i}x_{j+i}$, $i\leqslant j,\ j+i\leqslant(n-1)$; *and* 0 *otherwise,*

(iii) *If* $k>\frac{n}{2}$, *then* $H^*(V_{n,k};\mathbb{Z}_2)\cong\Lambda_{\mathbb{Z}_2}[x_k,x_{k+1},\ldots,x_{n-1}]$ *degree* $x_j=j$ $\mathrm{Sq}^i x_j=\binom{j}{i}x_{j+i}$, $i\leqslant j,\ j+i\leqslant(n-1)$; *and* 0 *otherwise.*

The above proposition can be proved via the transgression theorem of A. Borel and using the formulae of Proposition (1.1). We refer to [B 4] for such a proof. The above proposition can also be proved by using the nice cellular decomposition of Stiefel manifolds given by C. E. Miller. We refer to [Ch. IV of S 12] for the second proof.

Graphically, the above mod 2 cohomology algebras together with square operations can be neatly recorded in a diagram as follows: The diagram of $V_{n,k}$, denoted by $D(V_{n,k})$ (resp. $D(W_{n,k})$, $D(X_{n,k})$ for the diagram of $W_{n,k}$ and $X_{n,k}$) consists of $(n-k)$ points indexed by $k,k+1,\ldots,(n-1)$ and those bonds linking those pairs (i,j), $i<j$, with $\binom{i}{j-i}\not\equiv 0$ mod 2. We shall say that $V_{n,k}$ (resp. $W_{n,k}$; $X_{n,k}$) is *squarely connected* if the above *diagram*, $D(V_{n,k})$, *is connected.* As a preliminary indication of the type of results one may expect for transformation groups on Stiefel manifolds, let us prove the following theorem. The dimensional condition is unnecessarily restrictive, but is such that it will considerably simplify the proof.

Theorem (VII.4). *Let* $W_{n,k}$ *(resp.* $X_{n,k}$*) be complex (resp. quanternionic) Stiefel manifolds with* $k>\frac{1}{2}(n-3)$ *and connected diagram* $D(W_{n,k})$ *(resp.* $D(X_{n,k})$*). Let* $G=\mathrm{SO}(4m)$, *or* $\mathrm{SU}(2m)$, *or* $\mathrm{Sp}(m)$ *with*

$$m=2^b\geqslant\tfrac{1}{2}(n-k)\quad (resp.\ \geqslant(n-k))\,.$$

Suppose G acts on $W_{n,k}$ (resp. $X_{n,k}$) with non-empty fixed point set F and $\Omega_F=\{\pm\theta_i$, with multi. $l\}$. Then

(i) *l is a multiple of $(n-k)$ if $G=\mathrm{SO}(4m)$ or $\mathrm{SU}(2m)$; l is a multiple of $2(n-k)$ if $G=\mathrm{Sp}(m)$.*

(ii) *The fixed point set F has the same $\mathbb{Z}_2$-cohomology type of a suitable complex (resp. quanternionic) sub-Stiefel manifold and moreover $D(F)\cong D(W_{n,k})$ (resp. $D(X_{n,k})$).*

Proof. Since the proofs of the above six cases (i.e., two choices for the space and three choices for the group G) are essentially parallel, we shall only give the proof of the case $\mathrm{SU}(2m)$ acting on $W_{n,k}$ as follows:

(i) It follows from the condition $k\geqslant\frac{1}{2}(n-3)$ that $W_{n,k}$ satisfies the dimensional restriction of Theorem (VII.2′), i.e., $\mathrm{Max}\{(m_j+1)\}\leqslant\mathrm{Min}\{(m_i+m_j)\}$. Hence Theorem (VII.2′) applies to show that all isotropy subgroups, G_x, are of the form $\mathrm{SU}(m_x)$, and consequently,

$$F(T,X)=F(\mathbb{Z}_2^{2m-1},X)=F(G,X)=F$$

where T is a maximal torus and $\mathbb{Z}_2^{2m-1}$ is a maximal $\mathbb{Z}_2$-torus of G. Moreover, it follows easily from Theorem (IV.1) that F is also of the integral cohomology type of a product of $(n-k)$ odd spheres, namely

$$H^*(F,\mathbb{Z})\cong\Lambda_{\mathbb{Z}}[f_1,\dots,f_{(n-k)}]\,.$$

(ii) Since $F=F(G,X)=F(\mathbb{Z}_2^{2m-1},X)$ and $H^*(F:\mathbb{Z}_2)\cong\Lambda_{\mathbb{Z}_2}[f_1,\dots,f_{(n-k)}]$, it is clear that X is totally non-homologous to zero in the mod 2 Serre spectral sequence of the fibration $X_G\to B_G$ and

$$0\longrightarrow H_G^*(X:\mathbb{Z}_2)\xrightarrow{\iota^*}H_G^*(F:\mathbb{Z}_2)\longrightarrow H_G^*(X,F,\mathbb{Z}_2)\longrightarrow 0$$

is a short exact sequence. Let $x_1,x_2,\dots,x_{(n-k)}$ be the exterior generators $H^*(X,\mathbb{Z}_2)$, $\deg(x_j)=2(k+j)-1$, and $\bar{x}_j$ be a lifting of x_j in $H_G^*(X:\mathbb{Z}_2)$. Let

$$\iota^*(\bar{x}_j)=\textstyle\sum_{i=1}^{(n-k)}a_{ji}f_i+\text{higher exterior terms}\in H_G^*(F;\mathbb{Z}_2)$$

where $a_{ji}\in H^*(B_G;\mathbb{Z}_2)=\mathbb{Z}_2[c_2,\dots,c_{2m}]$. Then $f_1\wedge f_2\wedge\cdots\wedge f_{(n-k)}$ is the fundamental cohomology class of $F,\bar{x}_1\wedge\bar{x}_2\wedge\cdots\wedge\bar{x}_{(n-k)}$ is a *lifting* of the fundamental cohomology class of $X=W_{n,k}$ and it follows from

$$\iota^*(\bar{x}_1\wedge\bar{x}_2\wedge\cdots\wedge\bar{x}_{(n-k)})=\det(a_{ji})\,f_1\wedge f_2\wedge\cdots\wedge f_{(n-k)}$$

and Theorem (IV.10) that $\det(a_{ji})=c_{2m}^l$. Observe that $c_2,c_3,\dots,c_{2m}$ are algebraically independent. Hence, one may suppress $c_2,c_3,\dots,c_{(2m-1)}$ in the computation of $\det(a_{ji})$. Hence, letting $\bar{a}_{ji}$ be the polynomial of c_{2m} alone, obtained by substituting 0 for $c_2,c_3,\dots,c_{(2m-1)}$ into a_{ji}, we have $\det(\bar{a}_{ji})=c_{2m}^l$. Since a_{ji} is, by definition, homogeneous, $\bar{a}_{ji}$ is either 0 or is a power of c_{2m}. Therefore, $\bar{a}_{ji}\neq 0$ implies that $\deg\bar{x}_j\equiv\deg f_i \pmod{4m}$. On the other hand, it follows from $\det(\bar{a}_{ji})=c_{2m}^l\neq 0$ that, for each $1\leqslant j\leqslant(n-k)$, there is at least one i such that $\bar{a}_{ji}\neq 0$, i.e.,

$\deg \bar{x}_j \equiv \deg f_i \pmod{4m}$. Since it is assumed that $4m \geqslant 2(n-k)$, and $\deg \bar{x}_j = 2(k+j)-1$ for $1 \leqslant j \leqslant (n-k)$, it is clear that $\{\deg \bar{x}_j \pmod{4m};\ 1 \leqslant j \leqslant (n-k)\}$ are all distinct. Therefore, there must be a one-to-one correspondence between $\{\bar{x}_j\}$ and $\{f_i\}$ such that $\deg \bar{x}_j \equiv \deg f_i \pmod{4m}$. We shall assume that $\deg \bar{x}_j \equiv \deg f_j \pmod{4m}$, namely $\iota^*(\bar{x}_j) \equiv \bar{a}_{jj} f_j \bmod c_2, \ldots, c_{2m-1}$ and higher exterior terms where $\bar{a}_{jj} = c_{2m}^{l_j}$ and $\sum_{j=1}^{(n-k)} l_j = l$.

(iii) Next, we shall show that $l_1 = l_2 = \cdots = l_{(n-k)}$ and

$$\mathrm{Sq}^{2i} x_j = x_{j+i} \quad \text{if and only if} \quad \mathrm{Sq}^{2i} f_j = f_{j+i}.$$

Notice that $2m = 2^{(b+1)}$ implies that

$$\mathrm{Sq}^{2i} c_j \equiv 0 \qquad (\mathrm{mod}\ c_2, \ldots, c_{2m-1})$$

for all j and $i < 2m$. Moreover, it follows easily from the condition $k > \frac{1}{2}(n-3)$ that the lifting $\bar{x}_j$ of x_j into $H_G^*(X; \mathbb{Z}_2)$ is *unique modulo* $(c_2, \ldots, c_{2m-1})$. Hence

$$\mathrm{Sq}^{2i} x_j = x_{j+i} \quad \text{if and only if} \quad \mathrm{Sq}^{2i} \bar{x}_j \equiv \bar{x}_{j+i} \bmod (c_2, \ldots, c_{2m-1}),$$

and consequently,

$$\mathrm{Sq}^{2i} x_j = x_{j+i} \quad \text{if and only if} \quad \iota^* \mathrm{Sq}^{2i} \bar{x}_j \equiv \iota^* \bar{x}_{j+i} \bmod (c_2, \ldots, c_{2m-1})$$

i.e.,

$$\mathrm{Sq}^{2i} \iota^* \bar{x}_j \equiv \mathrm{Sq}^{2i} \bar{a}_{jj} f_j \equiv \bar{a}_{jj} \mathrm{Sq}^{2i} f_j \equiv \iota^* \bar{x}_{j+i} \equiv \bar{a}_{j+i,j+i} f_{j+i} \qquad (\mathrm{mod}\ c_2, \ldots, c_{2n-1}).$$

Therefore

$$\mathrm{Sq}^{2i} x_j = x_{j+i} \quad \text{if and only if} \quad \bar{a}_{j,j} = \bar{a}_{j+i,j+i} \quad \text{and} \quad \mathrm{Sq}^{2i} f_j = f_{j+i}.$$

Finally it follows from the assumption that $D(W_{n,k})$ is connected and $\bar{a}_{1,1} = \bar{a}_{2,2} = \cdots = \bar{a}_{(n-k),(n-k)}$, so that

$$\det(\bar{a}_{ji}) = \left(c_{2m}^{l_1}\right)^{(n-k)} = c_{2m}^{l}, \quad \text{i.e.,} \quad l = l_1 \cdot (n-k).$$

Then, it is not difficult to check that $H^*(F, \mathbb{Z}_2) \cong H^*(W_{\bar{n},\bar{k}}; \mathbb{Z}_2)$ as an $\mathfrak{a}(2)$-algebra, where $\bar{n} = (n - 2l_1 \cdot m)$ and $\bar{k} = (k - 2l_1 \cdot m)$. □

Remark. If one restricts the $\mathrm{SU}(n)$ action on $W_{n,k}$ to the subgroup $G = \mathrm{SU}(2m)$ imbedded via the representation $l_1 \cdot \mu_{2m} + (n - 2l_1 m)\theta$, then the fixed point set F is actually the sub-Stiefel manifold $W_{\bar{n},\bar{k}}$ with $\bar{n} = (n - 2l_1 \cdot m)$ and $\bar{k} = (k - 2l_1 m)$.

Corollary 1. *Let* $X = W_{n,k}$ *(resp.* $X_{n,k}$*) with* $k > \frac{1}{2}(n-3)$ *and connected diagram* $D(W_{n,k})$ *(resp.* $D(X_{n,k})$*). Let* $G = \mathrm{SO}(m)$, *or* $\mathrm{SU}(m)$ *or* $\mathrm{Sp}(m)$ *with*

$$\left[\frac{m}{d}\right] \geqslant 2^b \geqslant \tfrac{1}{2}(n-k) \qquad (\textit{resp.} \geqslant (n-k))$$

where $d=4,2,1$ *respectively. Suppose* G *acts on* X *with non-empty fixed point set* F *and* $\Omega_F=\{\pm\theta_i$, *with multi.* $l\}$. *Then* l *is a multiple of* $(n-k)$ *if* $G=\mathrm{SO}(m)$ *or* $\mathrm{SU}(m)$, *and is a multiple of* $2(n-k)$ *if* $G=\mathrm{Sp}(m)$.

Proof. The above corollary follows from Theorem (VII.4) applying to the restricted action of $\mathrm{SO}(2^{b+2})$, or $\mathrm{SU}(2^{b+1})$, or $\mathrm{Sp}(2^b)$ respectively. □

Next let us consider the following classification problem:

Problem 8. Let $M=G/H$ be a given compact homogeneous space with G acting transitively and *effective* on M. Classify all *effective* differentiable G-actions on M up to conjugation in $\mathrm{Diff}(M)$.

For most cases, it seems reasonable to expect that the final answer of the above classification problem is such that the *original given* transitive G-action is the only possible effective G-action on M. We shall use Stiefel manifolds as our testing spaces for the above problem.

Theorem (VII.5). *Let* M *be a manifold of the same homotopy type of* $W_{n,k}$ *(resp.* $X_{n,k}$*),* $k\geq\frac{1}{2}(n-3)\geq 3$, *with a given (non-trivial) differentiable action of* $G=\mathrm{SU}(n)$ *(resp.* $\mathrm{Sp}(n)$*). Then all orbits are of the same type of a suitable Stiefel manifold* $W_{n,l}$ *(resp.* $X_{n,l}$*) with* $l\geq k$, *and the orbit space* M/G *is of the same homotopy type of* $W_{l,k}$ *(resp.* $X_{l,k}$*). In fact,* M *is the total space of a fibration* $W_{n,l}\to M\to M/G\approx W_{l,k}$ *(resp.* $X_{n,l}\to M\to M/G\approx X_{l,k}$*).*

Corollary 1. *Suppose furthermore that the diagram* $D(W_{n,k})$ *(resp.* $D(X_{n,k})$*) can not be decomposed into the disjoint union of* $D(W_{n,l})$ *and* $D(W_{l,k})$ *(resp.* $D(X_{n,l})$ *and* $D(X_{l,k})$*) for a suitable* $k<l<n$. *Then the manifold* M *must be* $W_{n,k}$ *(resp.* $X_{n,k}$*) itself and the only non-trivial* G*-action on* M *is the given transitive action.*

Corollary 2. *Suppose* M *is a manifold of the same homotopy type as* $W_{n,k}$ *(resp.* $X_{n,k}$*),* $k>\frac{1}{2}(n-3)\geq 3$, *and* $G=\mathrm{SU}(m)$ *(resp.* $\mathrm{Sp}(m)$*). If* $m>n$, *there exists no non-trivial* G*-action on* M.

Remarks. (i) The above theorem and corollaries also hold for the case of SO-actions on $V_{n,k}$, $k\geq\frac{1}{2}(n-3)\geq 3$. However, the proof is slightly more involved due to the $\mathbb{Z}_2$-torsion in the real case.

(ii) The dimensional restriction $k>\frac{1}{2}(n-3)$ can be considerably improved by a more elaborate proof then the simple proof given here.

Proof of Theorem (VII.5). (1) Let G/H be the principal orbit type of the given G-action on M. Then $\dim H\geq\dim \mathrm{SU}(k)$ (resp. $\mathrm{Sp}(k)$), $k\geq\frac{1}{2}(n-3)\geq 3$, and G/H must be stably parallelizable. It is rather tedious but straightforward to combine the knowledge about large dimensional subgroups of classical groups [D1] and the effective computation of characteristic classes of homogeneous spaces in terms of roots and weights [B 11, H 5] to show that Stiefel manifolds are the *only* such homogeneous spaces satisfying the above two conditions. Therefore, the principal orbit type $G/H=W_{n,l}$ (resp. $X_{n,l}$), $l\geq k$. [We refer the reader to [H 5, G 6] for such a proof.]

(2) Applying Theorem (VII.2′) to the restricted $\mathrm{SU}(l)$ (resp. $\mathrm{Sp}(l)$) action on M, it is not difficult to see that all orbits are again Stiefel manifolds, namely, $G(x)=W_{n,l_x}$ (resp. X_{n,l_x}) $l_x\geq l$ for all $x\in M$. We claim that all orbits are in fact principal

orbits, i.e., $l_x = l$ for all $x \in M$. Suppose the contrary that there exist some $x \in M$ with $G_x = \mathrm{SU}(l_x)$, $l_x > l$. Then, it is easy to see that the slice representation φ_x of $\mathrm{SU}(l_x)$ at x is given by $\varphi_x = (l_x - l)\mu_{l_x} + m \cdot \theta$, where μ_{l_x} is the standard representation of $\mathrm{SU}(l_x)$ on $\mathbb{R}^{2l_x}$ and θ is the trivial one-dimensional representation. Therefore, the orbit space M/G is naturally a manifold with the image of principal orbits as interior points and the image of those singular orbits as the boundary points (which is by assumption *non-empty*). Moreover, $\dim M/G = \dim M - \dim$ (Principal orbits) = $\dim W_{l,k}$ (resp. $\dim X_{l,k}$). Since

$$\tilde{H}^j(G(x); k) = 0 \quad \text{for} \quad j \leqslant 2l \quad (\text{resp. } (4l+2)) \text{ and all} \quad x \in M,$$

it follows from Vietoris-Beagle mapping theorem that

$$\pi: M \to M/G$$

induces cohomology isomorphism $H^j(M/G; k) \xrightarrow{\cong} H^j(M; k)$ for $j \leqslant 2l$ (resp. $j \leqslant 4l+2$). Hence, there exist cohomology classes

$$\bar{y}_{2i+1} \in H^{2i+1}(M/G; k) \qquad (\text{resp. } \bar{y}_{4i+3} \in H^{4i+3}(M/G; k)) \qquad k \leqslant i < l,$$

such that the $\pi^*(\bar{y}_{2i+1})$ (resp. $\pi^*(\bar{y}_{4i+3})$) are non-zero and form a part of the exterior generators of $H^*(M; k)$. In particular,

$$\pi^*(\bar{y}_{2k+1} \cup \cdots \cup \bar{y}_{2l-1}) = y_{2k+1} \cup \cdots \cup y_{2l-1} \neq 0$$
$$(\text{resp. } \pi^*(\bar{y}_{4k+3} \cup \cdots \cup \bar{y}_{4l-1}) = y_{4k+3} \cup \cdots \cup y_{4l-1} \neq 0)$$

which is a contradiction because M/G is a compact manifold with *non-empty boundary* and hence its *top dimensional cohomology* is zero. (Notice that $\bar{y}_{2k+1} \cup \cdots \cup \bar{y}_{2l-1}$ (resp. $\bar{y}_{4k+3} \cup \cdots \cup \bar{y}_{4l-1}$) is a class in the top dimensional cohomology of M/G.) The above contradiction proves that all orbits in M are of the type $W_{n,l}$ (resp. $X_{n,l}$).

(3) Hence, M is the total space of the following fibration:

$$W_{n,l} \to M \to M/G \qquad (\text{resp. } X_{n,l} \to M \to M/G).$$

We shall show that M/G is of the same homotopy type of $W_{l,k}$ (resp. $X_{l,k}$). Since M is simply connected and the fibre is connected, it is clear that M/G is also simply connected. Next, let us consider the following mappings:

$$W_{l,k} \xrightarrow{\subseteq} W_{n,k} \approx M \xrightarrow{\pi} M/G.$$

Then, it is not difficult to see the composition of above maps induces cohomology isomorphism $H^*(M/G; \mathbb{Z}) \to H^*(W_{l,k}; \mathbb{Z})$, and hence, it follows from Hurewicz theorem that $W_{l,k} \to M/G$ is a homotopy equivalence. □

In the extreme case of $k=0$, $W_{n,0} = \mathrm{SU}(n)$ and $X_{n,0} = \mathrm{Sp}(n)$ are the spaces of the groups themselves, the adjoint action becomes another possibility. In fact, for quite a few cases, it can be proved that the *adjoint action* and the *transitive action* are essentially the only possible non-trivial action of a simple compact connected Lie group G on the manifold itself.

Conjecture. Let G be a compact connected *simple* Lie group and Φ be a *non-transitive, non-trivial* action of G on a manifold M of the same homotopy type as itself. Then the orbit structure of Φ must be *cohomologically* identical with that of the *adjoint action*, namely,

(i) the principal orbit type of Φ is the flag manifold G/T,

(ii) $F(T,M)\sim_{\mathbb{Z}} T$ and the weight system of local representation of T at $F(T,X)$ is given by $\Omega_F(\Phi)=\Delta(G)$ (the root system of G),

(iii) the Weyl group W acts on $F(T,M)$ as a differentiable transformation group generated by reflections and the orbit space $M/G=F(T,M)/W$ which is of the same cohomology type of the Cartan polyhedron G/Ad as stratified spaces.

(D) *High Order Borel Formulae*

For topological actions of elementary abelian groups on homology spheres, the important Borel formula reads as follows (cf. Th. (IV.7)):

$$(\dim X-\dim F)=\sum(\dim Y_i-\dim F)$$

where Y_i runs through all corank one F^0-varieties (resp. F-varieties) of X. In particular, if $F=F(T,X)=\emptyset$ is empty, the above formula becomes $(\dim X+1)=\sum(\dim Y_i+1)$. In the thesis of Golber [G 4], the following second order formula was proved for actions of elementary abelian groups on product of two odd spheres.

Theorem (VII.6) [D. Golber]. *Suppose that a torus T acts topologically on $X\sim S^p\times S^q$, p and q odd and suppose that $F(T,X)=\emptyset$. Then*

$$\begin{aligned} e(X)-eF(T)-\textstyle\sum_H[eF(H)-eF(T)]=\textstyle\sum_K\{&eF(K)-eF(T)\\ &-\textstyle\sum_{H\supseteq K}[eF(H)-eF(T)]\}\end{aligned}$$

where H runs through the subtori of corank one and K runs through the subtori of corank two in T and $eY=(m+1)(n+1)$ if $Y\sim S^m\times S^n$.

Remark. The assumption $F(T,X)=\emptyset$ is unnecessary. The same formula was proved later for the case $F(T,X)\neq\emptyset$ by T. Chang and T. Skjelbred [C 2].

In general, suppose that a torus T acts topologically on a space X of the rational homotopy type of a product of l odd spheres. Then, it follows from Theorem (VII.1) that there always exist some F^0-varieties of corank *at most* l. In the simplest extreme case that all F^0-varieties of X are of corank *at least* l, we have the following l-th order formula of Borel-Golber type:

Theorem (VII.7). *Let X be a space of the rational homotopy type of odd spheres with a given action of torus T. Suppose the coranks of all F^0-varieties in X are* *at least l, i.e., corank* $(T_x) \geqslant l$ *for all* $x \in X$. *Then, all* F^0*-varieties in X are again of the same homotopy type of product of l odd spheres and*

$$e(X) = \sum e(Y_i)$$

where Y_i *runs through all* F^0*-varieties of corank l in X and e(Z) is defined to be* $\Pi(n_j+1)$ *if* $Z \sim S^{n_1} \times \cdots \times S^{n_l}$.

Proof of Theorem (VII.7). (1) Let $\{L(X_T), d\}$ be the associated graded differential algebra of $X_T: L(X_T) = \Lambda[x_1, \ldots, x_l] \otimes R_T$ is an exterior algebra over R_T and the differential is given by

$$dx_i = a_i + \text{terms involving } x\text{'s} \qquad (1 \leqslant i \leqslant l) \qquad a_i \in R_T .$$

In view of the assumption that corank $(T_x) \geqslant l$ for all $x \in X$, it follows from Lemma (1.1) and Theorem (IV.6) that the *varieties* of the ideal $I = \langle a_1, \ldots, a_l \rangle$ are all linear and of codimension exactly l. Therefore, it follows from Macaulay's Theorem [p. 208, vol. II, Z1] that $\operatorname{codim} \langle a_{i_1}, a_{i_2}, \ldots, a_{i_j} \rangle = j$ for all $(i_1, \ldots, i_j)$ and $\langle a_{i_1}, a_{i_2}, \ldots, a_{i_j} \rangle : \langle a_{i_{j+1}} \rangle = \langle a_{i_1}, a_{i_2}, \ldots, a_{i_j} \rangle$ for all $(i_1, \ldots, i_{j+1})$.

(2) Assume that $\dim x_1 \leqslant \dim x_2 \leqslant \cdots \leqslant \dim x_l$. Let $F^i L(X_T)$ be the sub-exterior algebra (over R_T) generated by $\{x_1, \ldots, x_i\}$. Then

$$0 \subseteq F^1 L(X_T) \subseteq F^2 L(X_T) \subseteq \cdots \subseteq F^i L(X_T) \subseteq \cdots \subseteq F^l L(X_T) = L(X_T)$$

is an increasing filtration of $L(X_T)$ by graded differential subalgebra over R_T. Therefore, there is an associated spectral sequence converging to $H(L(X_T), d) = H^*(X_T)$. Straightforward computation taking into account of the above extremely nice ideal theoretical conditions will show that $H^*(X_T) \cong R_T/I$ as an R_T-algebra. Moreover, it is possible to change the exterior basis $\{x_1, x_2, \ldots, x_l\}$ of $L(X_T)$ into $\{\bar{x}_1, \bar{x}_2, \ldots, \bar{x}_l\}$ so that $d\bar{x}_i = a_i \in R_T$ for all $1 \leqslant i \leqslant l$. The reason is as follows: Suppose x_i is the smallest i such that $dx_i = a_i + g_i(x)$ with $g_i(x) \neq 0$. Then, $dg_i(x) = d^2 x_i - da_i = 0$. But we have already proved that $H^*(X_T) \cong R_T/I$, therefore all cycles of non-zero exterior degrees in $L(X_T)$ must be boundaries, i.e., $g_i(x) = df_i(x)$. Hence we may simply change x_i into $\bar{x}_i = (x_i - f_i(x))$. Keep going, and we get a new exterior basis $\{\bar{x}_1, \bar{x}_2, \ldots, \bar{x}_l\}$ with $d\bar{x}_i = a_i$.

(3) As a consequence of the above fact $d\bar{x}_i = a_i$, it follows easily from the localization theorem and Theorem (IV.I) that all F^0-varieties in X are again of the rational homotopy types of a product of l odd spheres. Let $P_{X_T}(t)$ be the Poincaré polynomial of $H^*(X_T, k)$. Then it is not difficult to use the nice ideal theoretical conditions of (1) to show that

$$P_{X_T}(t) = (1 - t^2)^{-r} \cdot \prod_{i=1}^{l} (1 - t^{(q_i + 1)})$$

where $r = rk(T)$ and $X \sim S^{q_1} \times \cdots \times S^{q_l}$.

The same computation applies to a corank l F^0-variety Y_j will give

$$P(Y_j)_T(t) = (1 - t^2)^{-r} \prod_{i=1}^{l} (1 - t^{(q_{ji} + 1)})$$

where $Y_j \sim S^{q_{j1}} \times \cdots \times S^{q_{jl}}$. Hence

$$\lim_{t \to 1} (1-t)^{(r-l)} \cdot P_{X_T}(t) = \frac{1}{2^r} \cdot \prod_{i=1}^{l} (q_i+1) = \frac{1}{2^r} e(X),$$

$$\lim_{t \to 1} (1-t)^{(r-l)} \cdot P_{(Y_j)_T}(t) = \frac{1}{2^r} \cdot \prod_{i=1}^{l} (q_{ji}+1) = \frac{1}{2^r} e(Y_j).$$

Therefore, the formula

$$e(X) = \sum e(Y_j)$$

follows directly from Proposition 1 of [§ 2, H 17]. □

In view of the above results, it is natural to ask the following:

Problem 9. What is the general form of an l-th order Borel formula for torus actions on a product of l odd spheres? Precisely, we want a formula which expresses $e(X)$ in terms of $e(Y)$ for those F^0-varieties, Y, of corank $\leqslant l$.

§ 2. Degree of Symmetry of Compact Homogeneous Spaces

For a given differentiable manifold M, the degree of (differentiable) symmetry of M, denoted by $N_d(M)$, is defined to be

$$N_d(M) = \operatorname{Max}\{\dim G;\ \text{all compact subgroups } G \subseteq \operatorname{Diff}(M)\}.$$

In the case of compact manifolds, the above notion of degree of symmetry is *identical* to the following invariant defined in terms of the symmetries of all compatible Riemannian metrics. Namely

$$N'_d(M) = \operatorname{Max}\{\dim ISO(v)\}$$

where v runs through all compatible Riemannian metrics on M.

The fact that $N_d(M) = N'_d(M)$ follows from the following well-known results: (i) the full isometry group of a Riemannian metric, v, on a compact manifold M is always compact, (ii) every compact subgroup $G \subseteq \operatorname{Diff}(M)$ can be realized as an isometry group of a suitable Riemannian metric on M. In this section we shall investigate the degree of symmetry of compact homogeneous manifolds and several related problems. Our central theme is to verify several testing cases of the following general conjecture:

Conjecture. Suppose $M = G/H$ is a compact homogeneous space and G is one of the *highest dimensional, effective, transitive, compact* transformation groups on M. Then $N_d(M) = \dim G$.

Observe that on the one hand, the totality of compact homogeneous spaces encompasses great many varieites of topological types, and on the other hand, the behavior of *transformation groups on a given space* is obviously very much influenced by the topological type of the given space. Hence, it is unlikely that the above conjecture will be ever susceptible to a uniform proof. In fact, the necessity to treat *spaces of different topological type separately* is, probably, one the build-in characteristics of the subject of topological transformation groups. Roughly speaking, the difficulty of the proof of the above conjecture is proportional to the complexity of the topological structure of G/H and is inverse proportional to the ratio $\dim G/\dim(G/H)$. Technically, one of the main difficulties in the computation of degree of symmetry of a given manifold, $N_d(M)$, lies in the fact that the *group type* of those highest dimensional transformation groups G on M, i.e., $\dim G = N_d(M)$, remains *undetermined* until the very end of the proof. One way of avoiding this difficulty is to consider the simple degree of symmetry, denoted by $\mathrm{SN}_d(M)$, which is by definition:

$$\mathrm{SN}_d(M) = \mathrm{Max}\{\dim G\}$$

where G runs through all *simple* compact Lie subgroups of $\mathrm{Diff}(M)$.

(A) *Degree of Symmetry of Compact Irreducible Symmetric Spaces of Lower Rank*

From a geometric viewpoint, those compact symmetric spaces of rank one are about the most homogeneous geometric structures. They are exactly the so-called two-point homogeneous spaces [W 1]. Hence, it is natural to expect that the natural homogeneous metric on such a space is indeed the most symmetric one. The cases $M = S^r$ or RP^n are classically well-known (Fubini-Birkhoff Theorem). The remaining cases of CP^n, QP^n and $F_4/\mathrm{Spin}(9)$ can be proved by cohomology method. We state the result as follows:

Theorem (VII.8) [H 18]. *Let (M, v_0) be a compact Riemannian symmetric space of rank one and M be the underlying differentiable structure. If v' is any other Riemannian metric on M, then*

$$\dim ISO(v') \leqslant \dim ISO(v_0) = N_d(M)$$

and the equality holds only when v' and v_0 are proportional.

Proof. The case $M = S^n$ or RP^n is well-known and easy. We shall prove the remaining cases of CP^n, QP^n and $F_4/\mathrm{Spin}(9)$ separately.

(1) *CP^n-case.* Let G be a compact Lie group acting effectively on CP^n and $\dim G \geqslant \dim \mathrm{SU}(n+1) = n^2 + 2n$. Let $T \subseteq G$ be a maximal torus of G and $\Omega = \{w_j, k_j\}$ be the weight system of the restricted T-action on CP^n. Since the G-action is assumed to be effective, it follows from Theorem (VI.1) that $\mathrm{Ker}(\Omega) = \{t \in T\colon w_1 = w_2 = \cdots = w_s\} = \{0\}$ and hence $\mathrm{rk}(T) \leqslant (s-1) \leqslant n$. Moreover, Ω is invariant under the action of Weyl group $W(G)$. Then it follows easily from the following table that $G = \mathrm{SU}(n+1)$ is the only possible compact Lie group of $\dim G \geqslant (n^2 + 2n)$, $\mathrm{rk}(G) \leqslant n$ and with an invariant (under Weyl group) weight system Ω, $\mathrm{Ker}(\Omega) = 0$.

Table 1. *Minimal Number of Weights and Dimensions of Simple Groups*

G	rk(G)	dim(G)	η_0 = Min number of invariant weights	dim G_{η_0}
A_r	r	r^2+2r	$(r+1)$	$(r+1)-(1/(r+1))$
B_r	r	$2r^2+r$	$2r$	$r+\frac{1}{2}$
C_r	r	$2r^2+r$	$2r$	$r+\frac{1}{2}$
D_r	r	$2r^2-r$	$2r$	$r-\frac{1}{2}$
G_2	2	14	6	$\frac{14}{6}$
F_4	4	52	24	$\frac{13}{6}$
E_6	6	78	27	$\frac{78}{27}$
E_7	7	133	56	$\frac{133}{56}$
E_8	8	246	238	$\frac{246}{238}$

(2) QP^n-*case.* Let G be a compact Lie group acting effectively on QP^n and $\dim G \geqslant \dim \mathrm{Sp}(n+1)=(2n^2+5n+3)$. Again, it follows from Theorem (VI.6) that $\mathrm{rk}(G) \leqslant (n+1)$ and the weight system Ω must be *even* and invariant under the Weyl group $W(G)$. Therefore, it is not difficult to read off from the above table that the only possible compact Lie groups satisfying the conditions: $\dim G \geqslant (2n^2+5n+3)$, $\mathrm{rk}(G) \leqslant (n+1)$ and with an *even*, invariant weight system Ω, $\mathrm{Ker}(\Omega)=\{0\}$, are C_{n+1} and B_{n+1}. Moreover, it is also easy to see that the G action must be transitive. However, B_{n+1} can not act transitively on QP^n. Hence $G=\mathrm{Sp}(n+1)$ and the invariant metric v' must be proportional to the symmetric metric v_0.

(3) $F_4/\mathrm{Spin}(9)$ *case.* Let G be any compact Lie group acting effectively on a manifold of the cohomology type of the Cayley projective plane. We claim that $\mathrm{rk}(G) \leqslant 4$, and hence $\dim G \leqslant \dim F_4$ (for F_4 is the highest dimensional Lie group of rank 4). To show that $\mathrm{rk}(G) \leqslant 4$, we may assume without loss of generality that $G=T$ is a torus. Then $H_T^*(X)=R_T[\zeta]/\langle f(\zeta)\rangle$ where $f(\zeta)=(\zeta-\alpha_1)(\zeta-\alpha_2)(\zeta-\alpha_3)$, $\alpha_i \in H^q(B_T)$. Suppose that $\alpha_1, \alpha_2, \alpha_3$ are distinct. Then $F(T, X)$ consists of three isolated fixed points. Moreover, simple computation will show that $(\alpha_1-\alpha_2)(\alpha_1-\alpha_3)$ splits into products of local weights at the fixed point p_1 (indexed by α_1) [by Theorem (IV.10)]. Hence there exists a corank two subtorus T' whose fixed point set $F(T', X)$ is connected and hence either of the type of CP^2 or QP^2 or the whole space. Suppose $F(T', X) \sim CP^2$. Then there exists a corank one subtorus of T', say T'', such that $F(T'', X) \sim QP^2$ or the whole space. Suppose $F(T'', X) \sim QP^2$, then one may restrict again to a suitable corank one subtorus of T'', say T''', with $F(T''', X)=X$. Namely, there exists a subtorus T''' of corank at most 4 acting trivially on X. But T is assumed to be effective, $\mathrm{rk}(T) \leqslant 4$. The other possibilities are easier and, in some cases, one can prove that $\mathrm{rk}(T)<4$. □

Next, let us consider those irreducible compact symmetric spaces of rank two. Among them, the group manifolds of simple compact Lie groups of rank two are topologically the simplest cases, namely, A_2, B_2 and G_2.

Theorem (VII.9). *Let M be the group manifolds of the simple compact Lie groups of rank two, i.e., A_2, B_2 or G_2. Then $N_d(M)=2\cdot\dim M$.*

The proofs of the three cases of A_2, B_2 and G_2 are essentially similar. Among them the case of G_2 is the most involved. We shall only give the proof of the G_2 case and leave the other two simpler cases to the reader. We need the following Proposition (2.1) which is rather interesting by itself.

Proposition (2.1). *There are only two possible (non-trivial) differentiable G_2-action on the group manifold of G_2 itself, namely, the transitive action and the adjoint action.*

Proof. (1) Let G_2/H be the principal orbit type of a given non-trivial G_2 action on itself. If $\dim H=0$, then the action is obviously transitive and $H=\{\mathrm{id}\}$. If $H=T$ is the maximal torus, then it is clear that $F=F(T,G_2)$ is a 2-torus (as a space) and the system of non-zero local weights of T around the fixed point set is given by $\Omega_F=\Delta(G_2)$. Furthermore, it is not difficult to show that the Weyl group $W(G_2)$ acts on F as a group generated by reflections and the orbit space of the given G_2 action is naturally identified with $F/W(G_2)$ which is a two simplex. From that, it is easy to see that such an action is equivalent to the adjoint action.

(2) Suppose that $\dim H>0$ and $H\neq T$. Since the group manifold G_2 is obviously a parallelizable manifold, it follows easily from the slice theorem that G_2/H is stably parallelizable. It is not difficult to check that $G_2/\mathrm{SU}(3)=S^6$, $G_2/\mathrm{SU}(2)=V_{7,5}$, and G_2/S (where S is a one dimensional subgroup of T) are the only stably parallelizable homogeneous spaces of G_2, besides the case G_2/T that has been already considered in (1). We shall show case by case that the above three cases are all *impossible.* The case G_2/S is easy to rule out. Since G_2/S is of codimension one, the orbit space must be an interval. Then it is quite straightforward to show that the manifold G_2 can not be built in this way. Next, suppose that the principal orbit type is $G_2/\mathrm{SU}(2)$. Then, it is not difficult to show, by the slice theorem, that $G_x^0=\mathrm{SU}(2)$, or $\mathrm{SU}(3)$ or G_2 and that the orbit space is a 3-dimensional rational cohomology manifold with *non-empty boundary.* Hence $H^3(X/G_2;k)=0$. On the other hand, it follows from the Vietoris-Beagle mapping theorem and the fact that $\tilde{H}^i(G(x);k)=0$ for $i\leqslant 4$ and all $x\in X$ that $H^3(X/G;k)\cong H^3(X;k)=k\neq 0$. The above contradiction proves that the principal orbit type can *not* be $G_2/\mathrm{SU}(2)$.

(3) Finally, we shall show that $S^6=G_2/\mathrm{SU}(3)$ can *not* be the principal orbit type either. Suppose the contrary. Then the only other possible orbit types are either a point or RP^6. If *there is no orbit of RP^6 type*, then it follows from the blowing-up argument of [H10] that the orbit space X/G is a manifold with the image of fixed points as boundary points. And moreover $X=\partial[(X/G)\times D^7]$. Then, simple topological consideration will show that the space of G_2 can not be obtained in this way (because $G_2\neq S^3\times S^{11}$).

Suppose there is at least one orbit, say $G(x_0)$, of RP^6 type. Let ξ be the canonical line bundle over RP^6. Then it follows from the slice theorem that the normal bundle $v(G(x_0))$ is of the form $a\xi+b\theta$ where $a+b=(14-6)=8$ and θ is the trivial line bundle. Moreover,

$$\tau(G(x_0))+\theta+v(G(x_0))=7\cdot\xi+a\xi+b\theta=15\cdot\theta.$$

Since the order of ξ in $\widetilde{KO}(RP^6)$ is 8, a must be equal to 1. Hence, the F-variety of x_0 is of the form of $RP^6\times M^7$ where M^7 is a *compact manifold without boundary*. If we restrict the G_2-action to one of its maximal $\mathbb{Z}_2$-torus say Q, it is clear that

$$\begin{aligned}\dim_{\mathbb{Z}_2} H^*(F(Q,X);\mathbb{Z}_2) &\geqslant \dim_{\mathbb{Z}_2} H^*(F(Q,RP^6\times M^7);\mathbb{Z}_2)\geqslant 7\cdot 2=14\\ &>8=\dim_{\mathbb{Z}_2} H^*(X;\mathbb{Z}_2)\end{aligned}$$

which is a contradiction to a result of A. Borel [B4].

All the above contradictions prove that the transitive action and the adjoint action are the only non-trivial G_2 actions on itself.

Corollary (2.1). *All differentiable action of* SO(n), $n\geqslant 7$, *on the manifold of* G_2 *must be trivial.*

Proof. It follows easily from Proposition (2.1) by restricting the given SO(n) action to $G_2\subseteq \mathrm{SO}(7)\subseteq \mathrm{SO}(n)$. □

Proof of Theorem (VII.9) for the G_2 case. Recall that

$$H^*(G_2;k)=\Lambda_k[x_3,x_{11}]\quad\text{and}\quad H^*(G_2;\mathbb{Z}_2)=\mathbb{Z}_2[x_3]/\langle x_3^4\rangle\otimes\Lambda_{\mathbb{Z}_2}[x_5],$$

where the suffix of each generator denotes the degree and $Sq^2x_3=x_5$ in $H^*(G_2;\mathbb{Z}_2)$. We refer to the papers of A. Borel [B4] and S. Araki [A5] for the computations of cohomologies of exceptional Lie groups. The crucial step in the proof is to show that there *does not* exist SO(5) actions on the space G_2 with $S^4=\mathrm{SO}(5)/\mathrm{SO}(4)$ as the principal orbit type.

(1) Suppose the contrary that there is such an SO(5)-action on the space of G_2 with S^4 as its principal orbit type. Again, we claim that *there is no orbit of RP^4-type*. The proof is essentially the same as that of Proposition (2.1). For otherwise, the F-variety of x_0 with $G(x_0)\sim RP^4$ must be of the following form

$$F(x_0)=RP^4\times M^j$$

where M^j is a compact manifold without boundary and of dimension $j=3$ or 7. Therefore, by restricting the action to a maximal $\mathbb{Z}_2$-torus Q,

$$\begin{aligned}\dim_{\mathbb{Z}_2} H^*(F(Q,G_2);\mathbb{Z}_2) &\geqslant \dim_{\mathbb{Z}_2} H^*(F(Q,RP^4\times M^j);\mathbb{Z}_2)\geqslant 5\cdot 2=10\\ &>\dim_{\mathbb{Z}_2} H^*(G_2;\mathbb{Z}_2)=8\end{aligned}$$

which contradicts a result of A. Borel. Hence, there is no orbits of RP^4-type and it follows from the blowing up argument of [H10] that the space G_2 equals $\partial(X'\times D^5)$ as an SO(5)-space, where X' is the orbit space with the image of fixed points as boundary points and the SO(5)-action on $X \times D^5$ is given by $g(x,d)=(x,g\cdot d)$. We claim that $\partial X' \neq \emptyset$. For otherwise $G_2 = X' \times S^4$ which is obviously impossible. Let $T^2 \subseteq \mathrm{SO}(4) \subseteq \mathrm{SO}(5)$ be a maximal torus of SO(5). Then $F(T^2, G_2) = F(T^2, \partial(X' \times D^5)) = \partial(X' \times D^1)$ = the doubling of X' along the boundary $\partial X'$ and $F(Q, G_2) = F(Q, \partial(X' \times D^5)) = \partial X'$. Observe that X' is 2-connected and $H^3(X', \mathbb{Z}) = H^3(G_2; \mathbb{Z}) = \mathbb{Z}$, and it is not difficult to show that $F(T^2, G_2) = F(\mathrm{SO}(4), G_2) = \partial(X' \times D^1) \sim_k S^3 \times S^7$. From that, it is easy to show that $\partial X' = F(Q, G_2)$ is connected and to deduce a contradiction from the mod 2 cohomology structure of G_2 and the following map

$$H^*_{\mathrm{SO}(5)}(G_2; \mathbb{Z}_2) \to H^*_{\mathrm{SO}(5)}(F; \mathbb{Z}_2) = H^*(\partial X'; \mathbb{Z}_2) \otimes H^*(B_{\mathrm{SO}(5)}; \mathbb{Z}_2).$$

(2) It follows easily from (1) that there is *no* action on the space of G_2 *with orbits solely of the type of* S^l, RP^l *or a point if* $l \geqslant 4$. Similarly, there is *no* SU(l)-action on G_2 with S^{2l-1} as principal orbit type if $l \geqslant 4$ and there is *no* Sp(l)-action on G_2 with S^{4l-1} as principal orbit type if $l \geqslant 2$. With all the above restrictions about the possibilities of principal orbit types of actions of simple compact Lie groups on the space of G_2, the proof of the G_2-case of Theorem (VII.9) is then but a straightforward enumeration. In fact, the only compact Lie group with dimension $\geqslant 28$ that can act effectively on the space of G_2 is $G_2 \times G_2$ acting via right and left translation. ▯

Next let us consider the second real Grassman manifolds $G_{n,2} = \mathrm{SO}(n)/\mathrm{SO}(n-2) \times \mathrm{SO}(2)$. In case n is odd, $G_{n,2}$ is of the same *rational* homotopy type of $CP^{(n-2)}$.

Theorem (VII.10). *Let* $G_{n,2} = \mathrm{SO}(n)/\mathrm{SO}(n-2) \times \mathrm{SO}(2)$ *and n odd. Then*

$$N_d(G_{n,2}) = \dim \mathrm{SO}(n) = \tfrac{1}{2} n(n-1).$$

Remark. Notice that $G_{n,2}$ and $CP^{(n-2)}$ are of the same rational homotopy type but $N_d(G_{n,2}) = \frac{1}{2} n(n-1)$ is only about the half of $N_d(CP^{(n-2)}) = (n^2 - 2n)$.

Proof. Recall that the integral and mod 2 cohomologies of $G_{n,2}$ are as follows. We refer to [B4] for their actual computation.

$$H^*(G_{n,2}; \mathbb{Z}_2) = \mathbb{Z}_2[u_2]/\langle u_2^r \rangle \otimes \Lambda_{\mathbb{Z}_2}[v_{(n-1)}],$$
$$H^*(G_{n,2}; \mathbb{Z}) = \mathbb{Z}[u_2, v_{(n-1)}]/\langle u_2^r - 2v_{(n-1)}; v_{(n-1)}^2 \rangle,$$

where $r = (n-1)/2$ and the suffix of each generator denotes the degree.

(1) If $l \geqslant (n+1)/2$, then there does not exist any non-trivial SU(l)-action on $G_{n,2}$. Suppose the contrary. Since $G_{n,2} \sim_k CP^{(n-2)}$, it follows from Theorem (IV.3) that

$$H^*_T(G_{n,2}; k) = R_T[\xi]/\langle f(\xi) \rangle$$

where $f(\xi) = (\xi - \alpha_1)^{k_1} \cdots (\xi - \alpha_s)^{k_s}$, $\sum k_k = (n-1)$, and the weight system $\Omega = \{\alpha_j, k_j\}$ is invariant under the Weyl group of SU(l). Due to the assumption $l \geqslant (n+1)/2$, it is clear that the only invariant weight pattern with $\sum k_j = (n-1)$ is the following

$$\Omega = \{\theta_1, \ldots, \theta_l \text{ with multi } 1, \text{ and } 0 \text{ with multi } (n-l-1)\}.$$

Then it is not difficult to see that there is an isolated orbit of the type of $CP^{(l-1)}$ such that $F(T, CP^{(l-1)})$ consists of those l isolated points of $F(T, G_{n,2})$ indexed by $\theta_1, \ldots, \theta_l$ respectively. By restricting the action to a maximal $\mathbb{Z}_2$-torus Q of $SU(l)$, it follows from the following commutative diagram that $\imath^*: H^*(G_{n,2}; \mathbb{Z}_2) \to H^*(CP^{(l-1)}; \mathbb{Z}_2)$ maps u_2 onto the generator $\eta \in H^2(CP^{(l-1)}; \mathbb{Z}_2)$: Namely,

$$\begin{array}{ccccc} H^*(G_{n,2}; \mathbb{Z}_2) & \longleftarrow & H_Q^*(G_{n,2}; \mathbb{Z}_2) & \longrightarrow & H_Q^*(F, \mathbb{Z}_2) \\ \big\downarrow & & \big\downarrow & & \big\downarrow \\ H^*(CP^{(l-1)}; \mathbb{Z}_2) & \longleftarrow & H_Q^*(CP^{(l-1)}; \mathbb{Z}_2) & \longrightarrow & H_Q^*(F(Q, CP^{(l-1)}); \mathbb{Z}_2). \end{array}$$

But this is a contradiction because $0 = \imath^*(u_2^r) = \imath^*(u_2)^r = \eta^r \neq 0$.

(2) Let G be an effective differentiable transformation compact connected Lie group on $G_{n,2}$ and T be a maximal torus of G. Let Ω be the weight system and G/H be the principal orbit type. Then Ω is invariant under the Weyl group $W(G)$ and the principal orbit type G/H can be computed from Ω (cf. Theorem (VI.2)). From that and the obvious fact $\operatorname{rk}(G) \leqslant (n-2)$, it is then not difficult to show by similar estimates as that of the CP^n-case in the proof of Theorem (VII.8) that $\dim G \leqslant \frac{1}{2}n(n-1)$. The only additional information that one needs is exactly the restriction that G contains no normal subgroup of the form $SU(l)$, $l \geqslant (n+1)/2$. □

(B) *Degree of Symmetry of Product of Spheres*

The following theorem was proved in [H 18]. We refer the reader to the original paper for its proof.

Theorem (VII.11) [H 18]. *Let M be an orientable manifold of dimension m and $H^*(M, k)$ be the rational cohomology algebra of M with compact support. If there exists a partition of $m, m = m_1 + m_2 + \cdots + m_l$ with each $m_j \geqslant 5$ and elements $\xi_j \in H^{m_j}(m, k)$ such that*

$$\xi_1 \cup \xi_2 \cup \cdots \cup \xi_l \neq 0 \quad in \quad H^m(M; k)$$

and moreover, the first rational Pontrjgin class of M vanishes, then

$$N_d(M) \leqslant N_d(S^{m_1} \times S^{m_2} \times \cdots \times S^{m_l}) = \sum \tfrac{1}{2} m_j(m_j + 1).$$

Corollary 1. *If $M^m = M_1^{m_1} \times M_2^{m_2} \times \cdots \times M_l^{m_l}$ it the product of l orientable compact manifolds with vanishing first Pontrjgin class, $m_j \geqslant 5$, then*

$$N(M_1^{m_1} \times M^{m_2} \times \cdots \times M^{m_l}) \leqslant N(S^{m_1} \times S^{m_2} \times \cdots \times S^{m_l}) = \sum \tfrac{1}{2} m_j(m_j + 1)$$

and equality holds only when $M_1^{m_1} \times M_2^{m_2} \times \cdots \times M_l^{m_l} \sim S^{m_1} \times S^{m_2} \times \cdots \times S^{m_l}$.

It was also conjectured in [H 18] that the above corollary should hold without the condition $P_1(M) = 0$ or $m_j \geqslant 5$.

(C) *Degree of Symmetry of Stiefel Manifolds*

It is well-known that Stiefel manifolds are of the rational homotopy type of product of spheres. However, except for a few cases where they are actually diffeomorphic to product of spheres such as $V_{q,6}=S^7\times S^6$, the degree of symmetry of a Stiefel manifold is usually much smaller than that of the product of spheres (of the same rational homotopy type). In fact, *if* the general conjecture for the degree of symmetry of compact homogeneous spaces turns out to be true for Stiefel manifolds, then, except for a few very special cases such as $V_{8,6}$, the degree of symmetry of Stiefel manifolds should be given by the following equalities:

$$\begin{aligned}
N_d(V_{n,k}) &= \dim(\mathrm{SO}(n)\times \mathrm{SO}(n-k)) = \tfrac{1}{2}n(n-1)+\tfrac{1}{2}(n-k)(n-k-1),\\
N_d(W_{n,k}) &= \dim(S(U(n)\times U(n-k))) = n^2+(n-k)^2-1,\\
N_d(X_{n,k}) &= \dim(\mathrm{Sp}(n)\times \mathrm{Sp}(n-k)) = 2n^2+n+2(n-k)^2+(n-k).
\end{aligned}$$

Under suitable dimensional restrictions such as $k\geqslant\frac{2}{3}n$ and connectedness of the associated diagram (defined in terms of the square operations of its mod 2 cohomology), the above equalities can actually be verified for most cases. [So far, the proofs are quite involved and ad hoc, and hence, are still unpublished.] We shall include here a proof of the following technically simpler special case:

Theorem (VII.12). *If* $n=(2^l+2^{l-1}-2)$ *and* $k=(2^l-1)$, *then*

$$\begin{aligned}
N_d(W_{n,k}) &= \dim(S(U(n)\times U(n-k))) = n^2+(n-k)^2-1, \quad \textit{and}\\
N_d(X_{n,k}) &= \dim(\mathrm{Sp}(n)\times \mathrm{Sp}(n-k)) = 2n^2+n+2(n-k)^2+(n-k).
\end{aligned}$$

Proof. The proofs of the cases of $W_{n,k}$ and $X_{n,k}$ are essentially parallel, we shall only give the proof of the $W_{n,k}$ case. Due to the assumption $n=(2^l+2^{l-1}-2)$ and $k=(2^l-1)$, the integral and mod 2 cohomology algebras of $W_{n,k}$ are as follows:

$$\begin{aligned}
H^*(W_{n,k};\mathbb{Z}) &= \Lambda_{\mathbb{Z}}[x_1,x_2,\dots,x_{(n-k)}],\\
H^*(W_{n,k};\mathbb{Z}_2) &= \Lambda_{\mathbb{Z}_2}[x_1,x_2,\dots,x_{(n-k)}],
\end{aligned}$$

where $\dim x_j=2(k+j)-1$, $1\leqslant j\leqslant(n-k)$. Moreover, the Steenrod square operations in $H^*(W_{n,k};\mathbb{Z}_2)$ are as follows (cf. Prop. (1.2), § 1—C):

$$\begin{aligned}
&\mathrm{Sq}^{2i}x_1 = x_{(i+1)}, \quad 1\leqslant i\leqslant(n-k-1)=(2^{l-1}-2), \quad\text{and}\\
&\mathrm{Sq}^{2i}x_j = x_{(i+j)} \quad \text{if} \quad (j-1)\equiv 0 \pmod{2^r}, \quad 1\leqslant i<2^r,\\
&\qquad\qquad\qquad \text{and} \quad (i+j)\leqslant(n-k).
\end{aligned}$$

Observe that $(n-k)=(2^{l-1}-1)$, $n=3(n-k)+1$, and

$$\begin{aligned}
\dim S(U(n)\times U(n-k)) &= n^2+(n-k)^2-1 = 10\cdot(n-k)^2+6(n-k),\\
\dim W_{n,k} &= n^2-k^2 = 5\cdot(n-k)^2+2(n-k).
\end{aligned}$$

That is, the *ratio* between $\dim S(U(n)\times U(n-k))$ and $\dim W_{n,k}$ is slightly bigger than 2.

(1) Suppose G is a compact connected Lie group acting almost effectively on $M=W_{n,k}$ and $\dim G>\dim S(U(n)\times U(n-k))=10\cdot(n-k)^2+6(n-k)$. Let G/H be the principal orbit type of such an action. Then it follows from the slice theorem and the parallelizability of $M=W_{n,k}$ that

(i) (G/H) is stably parallelizable,

(ii) $\dim G=a\cdot\dim G/H=a\cdot(\dim G-\dim H)$, $a>2+\dfrac{4}{5(n-k)+2}$, and

(iii) the isotropy representation $(\mathrm{Ad}_G|H-\mathrm{Ad}_H)$ is almost faithful.

Suppose G is non-simple and $G=G_1\times G_2$. Then H is locally isomorphic to $H_1\times H_d\times H_2$ where $H_1=G_1\cap H$, $H_2=G_2\cap H$ and H_d is the product of all simple normal subgroups of H such that

$$N\subseteq H\subseteq G\to G_i,\quad i=1,2$$

are both non-trivial (roughly speaking, N is diagronally embedded in $G_1\times G_2$). It follows easily from condition (ii), namely, $a>2$ and

$$\begin{aligned}\dim G=\dim G_1+\dim G_2&=a(\dim G-\dim H)\\ &=a(\dim G_1+\dim G_2-\dim H_1-\dim H_2-\dim H_d)\end{aligned}$$

that the following inequality holds at least for one factor, namely

$$\dim G_i+\dim H_d>a(\dim G_i-\dim H_i),\quad i=1 \text{ or } 2.$$

Observe that H_d is by definition locally isomorphic to a subgroup of

$$N(H_i,G_i)/H_i\quad(i=1,2),\quad\text{hence}\quad \dim H_d\leqslant\dim N(H_i,G_i)/H_i\quad(i=1,2).$$

It is not difficult to see that, by repeating the above argument if necessary, there exists at least one *simple normal* subgroup of G, say again denoted by G_1, such that

$$\dim G_1+\dim(N(H_1,G_1)/H_1)\geqslant a\cdot(\dim G_1-\dim H_1),\quad a>2+\frac{4}{5(n-k)+2}$$

where $H_1=G_1\cap H$ and G_1/H_1 is the principal orbit type of the restricted G_1-action on M. In this way, we reduce the proof of Theorem (VII.12) to showing that there *does not exist* any non-trivial action of *simple* compact Lie group *satisfying the above inequality.* We shall divide the proof of the above assertion into the following cases:

(2) *The case G_1 is an exceptional simple Lie group.* A detailed proof of this case involves tedious checking of high dimensional subgroups of exceptional Lie groups. In fact, such subgroups are rather scarce and it is quite straightforward to check that the only possible candidates of principal orbit types for

actions of exceptional compact Lie groups satisfying the above dimensional inequality are the following:

$$\begin{aligned}&G_2\colon\ S^6=G_2/\mathrm{SU}(3),\\ &F_4\colon\ F_4/\mathrm{Spin}(9),\quad F_4/\mathrm{Spin}(8),\\ &E_6\colon\ E_6/F_4,\\ &E_7\colon\ E_7/E_6\times T^1,\quad E_7/E_6,\quad E_7/D_6\times A_1,\\ &E_8\colon\ E_8/E_7\times A_1,\quad E_8/E_7.\end{aligned}$$

Among the above possibilities, most of them can be easily ruled out either because H is a maximal subgroup of G or because they are not stably parallelizable. The only cases that can not be ruled out by simple-minded slice considerations are $G_2/\mathrm{SU}(3)$ and $F_4/\mathrm{Spin}(8)$. However, G_2-actions with S^6 as principal orbit type can always be extended to SO(7)-actions with the same orbit structure, and F_4-actions with $F_4/\mathrm{Spin}(8)$, surely, restrict to a spin(8)-action with non-empty fixed point set. Therefore, it is convenient to take care of these two cases together with the classical groups.

(3) *The case* $G_1=A_{(m-1)}$. Let $G_1=\mathrm{SU}(m)$ and acting on $M=W_{n,k}$ with G_1/H_1 as its principal orbit type. We claim that it is *impossible* to have

$$\dim G_1+\dim(N(H_1,G_1)/H_1)\geqslant a(\dim G_1-\dim H_1),\quad a>2+\frac{4}{5(n-k)+2}.$$

Suppose the contrary. It follows from the above dimensional restriction and the stable parallelizability of G_1/H_1 that $H_1\sim\mathrm{SU}(m')$ with $m'>\frac{2}{3}m$. Applying Theorem (VII.2') to the restricted action of $\mathrm{SU}(m')$, it is easy to see that $F(\mathrm{SU}(m'),M)=F(T^{(m'-1)},M)\sim_{\mathbb{Z}}S^{d_1}\times\cdots\times S^{d_{(n-k)}}$, the local representation of $\mathrm{SU}(m')$ around $F(\mathrm{SU}(m'),M)$ is given by $(m-m')[\mu_m]_{\mathbb{R}}$ which is the real form of the usual $\mathrm{SU}(m')$ representation on $\mathbb{C}^{n'}$, and all isotropy subgroups of the $\mathrm{SU}(m')$-action are of the form of standardly imbedded $\mathrm{SU}(m_x)$. Since $m'>\frac{2}{3}m$, it is clear that $m'>2\cdot(m-m')$ and there exists a power of 2, say 2^r, such that

$$m'\geqslant 2^r>(m-m').$$

Now, we shall futher restrict the action to $G'=\mathrm{SU}(2^r)$. Then, again, all isotropy subgroups are of the form of standardly imbedded $\mathrm{SU}(m_x)$. Let $Q=\mathbb{Z}_2^{(2^r-1)}$ be a maximal $\mathbb{Z}_2$-torus of $\mathrm{SU}(2^r)$. It is clear that $F(Q,M)=F(\mathrm{SU}(2^r),M)$ is of the mod 2 comohology type of produkt of $(n-k)$ odd dimensional spheres, and

$$H^*_{G'}(M,\mathbb{Z}_2)\xrightarrow{\ i^*\ }H^*_{G'}(F(G',M);\mathbb{Z}_2)$$

is a monomorphism. Let $\bar{x}_1$ be the unique lifting of $x_1\in H^*(M;\mathbb{Z}_2)$ into $H^*_{G'}(M;\mathbb{Z}_2)$ which restricts to zero on fixed points, and let $\bar{x}_{j+1}=\mathrm{Sq}^{2^j}\bar{x}_1$ which is obviously a lifting of x_{j+1}. Then $H^*_{G'}(M;\mathbb{Z}_2)=\Lambda_{R_{G'}}[\bar{x}_1,\bar{x}_2,\ldots,\bar{x}_{n-k}]$

and $H^*_{G'}(F(G',M);\mathbb{Z}_2)=\Lambda_{R_{G'}}[f_1,f_2,\ldots,f_{n-k}]$, where $R_{G'}=H^*(B_{G'};\mathbb{Z}_2)$. Let $\imath^*(\bar{x}_j)=\sum a_{ij}f_i$. Then, it follows from Theorem (IV.10) that

$$\det(a_{ij})=C_{2^r}^{(m-m')}$$

where C_{2^r} is the mod 2 2^r-th Chern class in $R_{G'}=\mathbb{Z}_2[C_2,\ldots,C_{2^r}]$. Observe that $C_2,C_3,\ldots,C_{2^r}$ are algebraically independent. Hence one may suppress $C_2,C_3,\ldots,C_{(2^r-1)}$ in the computation of $\det(a_{ij})$. Namely, let $\bar{a}_{ij}$ be the monomial of C_{2^r} obtained by substituting 0 for $C_2,C_3,\ldots,C_{2^r-1}$ into a_{ij}, we have $\det(\bar{a}_{ij})=C_{2^r}^{(m-m')}$. Since we only need C_{2^r} for later computations, we shall denote C_{2^r} simply by C.

If $2^r\geqslant(n-k)$, then by Theorem (VII.4), $(m-m')$ is a multiple of $(n-k)$ and hence $\geqslant(n-k)$. However, $m>3\cdot(m-m')$ implies that $m>n$ and hence $\dim(G_1/H_1)=(m^2-m'^2)>\dim W_{n,k}$, which is obviously a contradiction. In case $2^r<(n-k)=2^{l-1}-1$, i.e., $r\leqslant(l-2)$, observe that $\bar{a}_{ji}\neq 0$ only when $\dim x_j\equiv\dim f_i \pmod{2^r}$, let $f_1, f_{2^r+1}, f_{2\cdot 2^r+1},\ldots$ be those exterior generators of $H^*(F(G',M);\mathbb{Z}_2)$ such that

$$\dim f_{b\cdot 2^r+1}\equiv\dim x_1 \pmod{2^r}.$$

Then, the determinant of the submatrix A_1 of $(\bar{a}_{ji})$ expressing $\imath^*(\bar{x}_1)$, $\imath^*(\bar{x}_{2^r+1}),\ldots$ in terms of $f_1, f_{2^r+1},\ldots \pmod{c_2,\ldots,c_{2^r-1}}$ must be non-zero. Similarly, let $f_j, f_{2^r+j},\ldots$ be those exterior generators of $H^*(F(G',M);\mathbb{Z}_2)$ such that

$$\dim f_{b\,2^r+j}\equiv\dim\bar{x}_j \pmod{2^r},\quad j=2,3,\ldots,2^r;$$

and A_j be the submatrix of $(\bar{a}_{ji})$ expressing $\imath^*(\bar{x}_j)$, $\imath^*(\bar{x}_{2^r+j}),\ldots$ in terms of $f_j, f_{2^r+j},\ldots \bmod(C_2,C_3,\ldots,C_{2^r-1})$. Therefore,

$$(\bar{a}_{ji})\sim\begin{pmatrix}A_1 & & & \\ & A_2 & & 0\\ 0 & & \ddots & \\ & & & A_{2^r}\end{pmatrix}\quad\text{and}\quad\det(\bar{a}_{ji})=\prod_{j=1}^{2^r}\det A_j.$$

Suppose $\bar{a}_{1i}\equiv 0 \pmod{C}$ for all i, then it follows easily from $\imath^*\bar{x}_{j+1}=\mathrm{Sq}^{2j}\imath^*\bar{x}_1$ that $\bar{a}_{ji}\equiv 0 \pmod{C}$ for all j and i and consequently $\det(\bar{a}_{ji})=C^{(m-m')}\equiv 0 \pmod{C^{(n-k)}}$, i.e., $(m-m')\geqslant(n-k)$, which is clearly a contradiction. On the other hand, suppose all the non-zero $\bar{a}_{1i}$ are equal to 1. Then it is easy to see that $\imath^*(\bar{x}_1)=f_1$ and it follows from the injectivity of $\imath^*$ that

$$\imath^*\bar{x}_{j+1}=\mathrm{Sq}^{2j}\imath^*x_1=\mathrm{Sq}^{2j}f_1\neq 0\quad\text{for all}\quad 1\leqslant j\leqslant(n-k-1)$$

and hence $\dim F(G',M)=\dim M$, which is obviously a contradiction because the G'-action is non-trivial. Therefore, the only remaining case is that

$$\imath^*x_1=f_1+\sum_{i<1}\bar{a}_{1i}f_i$$

with at least one non-zero $\bar{a}_{1i}$, $i>1$. Then, it follows from

$$\iota^* \bar{x}_{2^r+1} = \mathrm{Sq}^{2^{r+1}} \cdot (f_1 + \textstyle\sum_{i<1} \bar{a}_{1i} f_i), \ldots$$

and simple dimensional counting that

$$\textstyle\sum \dim f_{b\,2^r+1} \leqslant (\sum \dim x_{b\,2^r+1}) - 2 \cdot 2^r.$$

Hence $\det A_1 \equiv 0 \pmod{C^2}$. Moreover, it follows from the following fact:

$$\mathrm{Sq}^{2j}(x_{b \cdot 2^r+1}) = x_{b \cdot 2^r+j+1} \quad \text{for all} \quad 1 \leqslant j \leqslant (2^r-1)$$

that $\det A_j = \det A_1$ for all $1 \leqslant j \leqslant (2^r-1)$. Therefore

$$\det(\bar{a}_{ji}) = \textstyle\prod_{j=1}^{2^r} \det A_j \equiv 0 \pmod{C^{(2^{r+1}-2)}},$$

which is a contradiction to the fact $\det(\bar{a}_{ji}) = C^{(m-m')}$, $(m-m') < 2^r$. All the above contradictions prove that it is impossible to have $G_1 = \mathrm{SU}(m)$ acting on $M = W_{n,k}$ with G_1/H_1 satisfying

$$\dim G_1 + \dim(N(H_1, G_1)/H_1) \geqslant a(\dim G_1 - \dim H_1), \quad a > 2 + \frac{4}{5(n-k)+2}$$

as its principal orbit type.

The proofs for the cases $G_1 = \mathrm{Sp}(m)$ or $G_1 = \mathrm{SO}(m)$ are essentially parallel to that of the $G_1 = \mathrm{SU}(m)$ case, and hence are left to the reader. Finally, we remark that the above proof shows that the only effective compact transformation group on $W_{n,k}$ with dimension equal to the dimension of $S(U(n) \times U(n-k))$ is the obvious transitive transformation of $S(U(n) \times U(n-k))$ itself. □

Remark. Slight modifications of the above proof will show that

$$N_d(W_{n,k}) = \dim\big(S(U(n) \times U(n-k))\big),$$
$$N_d(X_{n,k}) = \dim(\mathrm{Sp}(n) \times \mathrm{Sp}(k))$$

for the case $k = (2^l - 1)$ and $n < (2^l + 2^{l-1} - 2)$. Similar proofs also work for real Stiefel manifolds. the 2-torsions of the cohomology of real Stiefel manifolds are, in fact, rather useful.

References

A 1 Adams, F.: On the non-existence of elements of Hopf invariant one. Ann. of Math. **72**, 20—104 (1960).

A 2 Adams, F.: Vector fields on spheres, Ann. of Math. **75**, 603—632 (1962).

A 3 Allday, C.: On the rank of a space. Trans. Amer. Math. Soc. **166**, 173—184 (1972).

A 4 Allday, C.: The stratification of compact connected Lie group actions by subtori (preprint, memeo. at Univ. Hawaii).

A 5 Araki, S.: Cohomology mod 2 of the compact exceptional groups E_6 and E_7. J. Math. Osaka City Univ. **12**, 43—65 (1961).

A 6 Atiyah, M.: Characters and cohomology of finite groups. Publ. Math. Inst. Hautes Etudes Scient. 1961.

A 7 Atiyah, M., Hirzebruch F.: Vector bundles and homogeneous spaces. Proc. of Symp. in Pure Math. **3**, Amer. Math. Soc. 1961

A 8 Atiyah, M., Segal, G.: Equivariant *K*-theory. Lecture notes, Warwick 1965.

B 1 Borel, A.: Some remarks about Lie groups transitive on spheres and tori. Bull. Amer. Math. Soc. **55**, 580—587 (1949).

B 2 Borel, A.: Sur la cohomologie des espaces fibrés principaux et des espaces homogènes de groupes de Lie compacts. Ann. of Math. **57**, 115—207 (1953).

B 3 Borel, A.: Les bouts des espaces homogenes des groupes de Lie. Ann. of Math. **58**, 443—457 (1953).

B 4 Borel, A.: La cohomologie mod 2 de certains espaces homogènes. Comment. Math. Helv. **27**, 165—197 (1953).

B 5 Borel, A.: Nouvelle démonstration d'un théorème de P. A. Smith. Comment. Math. Helv. **29**, 27—39 (1955).

B 6 Borel, A.: Transformation groups with two classes of orbits. Proc. Nat. Acad. Sci. U.S.A. **43**, 983—985 (1957).

B 7 Borel, A. Linear Algebraic Groups. N. Y.: Benjamin.

B 8 Borel, A.: Fixed points of elementary commutative groups. Bull. Amer. Math. Soc. **65**, 322—326 (1959).

B 9 Borel, A.: Le plan projectif de octaves et les spheres homogenes. Comptes Rendue de l'Academic des Sciences, Paris **230**, 1378—1383 (1960).

B 10 Borel, A. *et al.*: Seminar on Transformation Groups. Ann. of Math. Studies **46**, Princeton, N.J.: Princeton University Press 1961.

B 11 Borel, A., Hirzebruch, F.: Characteristic classes and homogeneous spaces I, Amer. J. Math. **80**, 485—538 (1958); II, **81**, 351—382 (1959); III, **82**, 491—504 (1960).

B 12 Borel, A., DeSiebenthal: Sur les sous-groupes fermes de rang maximum des groupes des Lie compacts connexes. Comm. Math. Helv. **23**, 200—221 (1949).

B 13 Borel, A., Serre, J. P.: Sur certain sous-groupes de Lie compacts. Comm. Math. Helv. **27**, 128—139 (1953).

B 14 Bott, R.: The stable homotopy of the classical groups. Ann. of Math. **70**, 313—337 (1959).

B 15 Bott, R. and Samelson: Application of the theory of Morse to symmetric spaces. Amer. J. Math. **80**, 964—1029 (1958).

B 16 Bredon, G.: The cohomology ring structure of a fixed set. Ann. of Math. **80**, 524—537 (1964).

B 17 Bredon, G.: On homogeneous cohomology spheres. Ann. of Math. **73**, 556—565 (1961).

B 18 Bredon B.: Transformation groups on spheres with two types of orbits. Topology **3**, 115—122 (1965).

B 19 Bredon, G.: Examples of differentiable group actions. Topology **3**, 103—113 (1965).

B 20 Bredon, G.: Cohomological aspects of transformation groups. In: Proc. Conf. on Transf. Groups, New Orleans, 1967, 245—280. Berlin-Heidelberg-New York: Springer 1968.

B 21 Brieskorn, E.: Beispiele zur Differentialtopologie von Singularitaten. Invent. Math. **2**, 1—14 (1966).

B 22 Browder, W., Livesay, G. R.: Fixed point free involutions on homotopy spheres. Bull. Amer. Math. Soc. **73**, 242—245 (1967).

B 23 Brown, E. Jr.: Cohomology theories. Ann. of Math. **75**, 467—484 (1962).

C 1 Chang, T.: Thesis. U. C. Berkeley (1972).

C 2 Chang, T., Skjelbred, T.: The topological Schur lemma and related results. Ann. of Math. **100**, 307—321 (1974).

C 3 Chevalley, C.; Theory of Lie groups. Princeton, N. J.; Princeton Univ. Press 1946.

C 4 Chevalley, C.: The Betti numbers of exceptional simple Lie groups. Proc. of International Congress of Math. **2**, pp. 21—24. Cambridge, USA (1950)

C 5 Conner, P.: Orbits of uniform dimension. Michigan Math. J. **6**, 25—32 (1959).

C 6 Conner, P., Floyd, E.: On the construction of periodic maps without fixed points. Proc. Amer. Math. Soc. 354—360 (1959).

C 7 Conner, P., Floyd, E.: Differentiable Periodic Maps. Berlin-Göttingen-Heidelberg-New York: Springer 1964.

C 8 Conner, P., Montgomery, D.: "An example for SO(3) action" Proceedings of the National Academy of Sciences, U.S.A., **48**, 1918—1922 (1962).

D 1 Dynkin, E.: Maximal subgroups of the classical groups. Trudy Moscov. Mat. Obšč., vol. **1**, 39—166 (1952); Amer. Math. Soc. Transl. Ser. 2, **6**, 245—378 (1957).

E 1 Eisenhart, L. P.: Riemannian Geometry. Princeton, N. J.: Princeton University Press 1940.

E 2 Eisenhart, L. P.: Continuous groups of transformations. Princeton, N. J.: Princeton University Press 1933.

E 3 Eilenberg, S., Moore, J. C.: Homology and fibrations I. Comment. Math. Helv. **40**, 199—236 (1966).

F 1 Floyd, E. E.: On periodic maps and the Euler characteristic of associated spaces. Trans. Amer. Math. Soc. **72**, 138—147 (1952).

F 2 Floyd, E. E.: Examples of fixed point sets of periodic maps I. Ann. of Math. **55**, 167—171 (1952); II, **64** 396—398.

F 3 Floyd, E. E.: Fixed point sets of compact abelian Lie groups of transformations. Ann. of Math. **66**, 30—35 (1957).

F 4 Floyd, E. E., Richardson, R.: An action of a finite group on an n-cell without stationary points. Bull. Amer. Math. Soc. **65**, 73—76 (1959).

G 1 Gleason, A.: Spaces with a compact Lie group of transformations. Proc. Amer. Math. Soc. **1**, 35—43 (1950).

G 2 Godement, R.: Theorie des faisceaux, Actualites Scientifiques et Industrielles, 1252. Paris: Hermann 1958.

G 3 Golber, D.: The cohomological description of a torus action. Pacific J. Math. **46**, 149—154 (1973).

G 4 Golber, D.: Torus actions on an product of two odd spheres. Topology **10**, 313—326 (1971)

G 5 Grothendieck, A.: Sur quelques points d'algebre homologique. Tôhoku Math. J. **9**, 119—221 (1957).

G 6 Grove, A.: SU(n) actions on manifolds with vanishing first and second integral Pontrjagin classes. In: Proc. 2nd Conf. on Compact Transformation Groups Amherst, Mass. 1971, Lecture Notes in Math. 298, 324—333, Berlin-Heidelberg-New York: Springer 1972.

H 1 Helgason, S., Differential Geometry and Symmetric Spaces. New York: Academic Press 1962.

H 2 Hirzebruch, F.: Singularities and exotic spheres. Seminaire Bourbaki 19 (1966/67).

H 3 Hirzebruch, F.: The topology of normal singularities of an algebraic surface (d'apres un article de D. Mumford). Seminaire Bourbaki 15 (1962/63).

H 4 Hsiang, W. C., Hsiang, W. Y.: Classification of differentiable actions on S^n, R^n and D^n with S^k as the principal orbit type. Ann. of Math. **82**, 421—433 (1965).

H5 Hsiang, W.C., Hsiang, W.Y.: Differentiable actions of compact connected classical group I. Amer. J. Math. **89**, 705—786 (1967). II, Ann. of Math. **92**, 189—223 (1970). III (to appear in Ann. of Math.).

H6 Hsiang, W.C., Hsiang, W.Y.: On compact subgroups of diffeomorphism groups of Kervaire spheres. Ann. of Math. **85**, 359—369 (1967).

H7 Hsiang, W.C., Hsiang, W.Y.: Some problems in differentiable transformation groups Proc. Conf. on Transf. Groups, New Orleans, 1967. Berlin-Heidelberg-New York: Springer 1968.

H8 Hsiang, W.C., Hsiang, W.Y.: A fixed point theorem for finite diffeomorphism groups generated by reflections, The Steenrod Algebra and Its Applications. Lecture Notes in Math. 168. Berlin-Heidelberg-New York: Springer 1970.

H9 Hsiang, W.C., Hsiang, W.Y.: Degree of symmetries of homotopy spheres. Ann. of Math. **89**, 52—67 (1969).

H10 Hsiang, W. Y.: On classification of differential $SO(n)$ actions on simply connected π-manifolds. Amer. J. Math. **88**, 137—153 (1966).

H11 Hsiang, W. Y.: On the bound of the dimensions of the isometry groups of all possible Riemannian metrics on an exotic sphere. Ann. of Math. **85**, 351—358.

H12 Hsiang, W. Y.: The natural metric on $SO(n)/SO(n-2)$ is the most symmetric metric. Bull. of Math. **73**, 55—58 (1967).

H13 Hsiang, W. Y.: On generalizations of a theorem of A. Borel and their applications in the study of topological actions. Proc. Topology Conf., Athens, Ga. 1969.

H14 Hsiang, W. Y.: On the geometric weight system of differentiable compact transformation groups on acyclic manifolds. Invent. Math. **12**, 35—47 (1971).

H15 Hsiang, W. Y.: On characteristic classes and the topological Schur lemma from the topological transformation groups viewpoint. Proc. Symposia in Pure Math. XXII. 105—112 (1971).

H16 Hsiang, W. Y.: On the splitting principle and the geometric weight system of topological transformation groups. I. In: Proc. 2nd Conf. on Compact Transf. Groups. Amherst. Mass. 1971. Lecture Notes in Math. 298. 334—402. Berlin-Heidelberg-New York: Springer 1972.

H17 Hsiang, W. Y.: On some fundamental theorems in cohomology theory of topological transformation groups. Taita (Nat'l. Taiwan Univ.) J. Math. **2**, 66—87 (1970) [some of the results were announced in Bull. Amer. Math. Soc **77**, 1096—1098 (1971)].

H18 Hsiang, W. Y.: On the degree of symmetry and the structure of highly symmetric manifolds. Tamkang J. of Math. (Taipei) **2**, 1—22 (1971).

H19 Hsiang, W. Y.: Structural theorems for topological actions of $\mathbb{Z}_2$-tori on real, complex and quaternionic projective spaces (to appear). Comm. Math. Helv. **49**, 479—491 (1974).

H20 Hsiang, W. Y., Su, J. C.: On the classification of transitive effective actions on Stiefel manifolds. Trans. Amer. Math. Soc. **130**, 322—336 (1968).

H21 Hsiang, W.Y., Su, J.C.: On the geometric weight system of topological actions on cohomology quaternionic projective spaces. Invent. Math. **28**, 107—127 (1975).

J1 Jacobson, N.: Lie Algebra. Interscience, 1962.

J2 Jänich, K.: Differenzierbare Mannigfaltigkeiten mit Rand als Orbitraume differenzierbarer G-Mannigfaltigkeiten ohne Rand. Topology **5**, 301—320 (1966).

K1 Kervaire, M.: A manifold which does not admit any differentiable structure. Comment. Math. Helv. **34**, 257—270 (1960).

K2 Kervaire, M., Milnor, J.: Groups of homotopy spheres I. Ann. of Math. **77**, 504—537 (1963).

K3 Koszul: Sur certains espaces de transformation de Lie. Strasbourg: Coll. de Géometrie Diff. 1953, C.N.R.S.

K4 Kramer, M.: Hauptisotropiegruppen bei endlich dimensionalen Darstellungen kompakter halbeinfacher Liegruppen (Diplomarbeit, Bonn 1966).

L1 Lefschetz, S.: Algebraic Topology, 1942.

L2 Liao, S. D.: A theorem on periodic transformations of homology spheres. Ann. of Math. **56**, 68—83 (1952).

M1 Massey, W.: Exact couples in algebraic topology, I, II. Ann. of Math. **56**, 363—396 (1952). III, IV, V Ann. of Math. **57**, 248—286 (1953)

M2 Milnor, J.: Construction of universal bundles. I, Ann. of Math. **63**, 272—284 (1956). II, Ann. of Math. **63**, 430—436 (1956).

M3 Montgomery, D. Samelson, H.: Transformation groups of spheres. Ann. Math. **44**, 454—470 (1943).

M 4 Montgomery, D., Yang, C. T.: The existence of slice. Ann. Math. **65**, 108—116 (1957).

M 5 Montgomery, D., Yang, C.T.: Differentiable actions on seven homotopy spheres. Transactions AMS **122**, 480—498 (1966).

M 6 Montgomery, D., Samelson, H., Yang, C. T.: Exceptional orbits of highest dimensions. Ann. of Math. **64**, 1—9 (1956)

M 7 Montgomery, D., Zippin, L.: Topological Transformation Groups. Interscience, 1955.

M 8 Mostow, G. D.: On a conjecture of Montgomery. Ann. Math. **65**, 432—446 (1957).

M 9 Mostow, G. D.: Equivariant embedding in euclidean space. Ann. Math. **65**, pp. 432—446 (1957).

M 10 Myers, S., Steenrod, N.: The group of isometries of a Riemannian manifold. Ann. of Math. **40**, 400—416 (1939).

O 1 Onscik, A.: Inclusion relations among transitive compact transformation groups. Trudy Moskov. Mat. Obšč. **11**, 199—242 (1962). (Translated in: Trans. Amer. Math. Soc. **50**, 2nd Series).

O 2 Onscik, A.: Transitive compact transformation groups. Mat. Sb. **60** (102), 447—485.

P 1 Palais, R.: Embedding of compact differentiable transformation groups in orthogonal representations. J. Math. Mech. **6**, 673—678 (1957).

P 2 Petrie, T.: S^1-actions on homotopy complex projective spaces. Bull. Amer. Math. Soc. **78**, 105—153 (1972).

P 3 Pontrjagin, L. *S.*: Topologische Gruppen. Leipzig: B. G. Teubner 1958.

Q 1 Quillen, D.: The spectrum of an equivariant cohomology ring: I, II. Ann. of Math. **94**, 549—602 (1971).

R 1 Richardson, M., Smith, P. A.: Periodic transformations on complexes. Ann. of Math. **39**, 611—633 (1938).

S 1 Samelson, H.: Topology of Lie groups. Bull. Amer. Math. Soc. **58**, 2—37 (1952).

S 2 Samelson, H.: Beitrage zur Topologie der Gruppenmannigfaltigkeiten. Ann. of Math. **42**, 1091—1137 (1941).

S 3 Serre, J. P.: Homologie singuliere des espaces fibres. Ann. of Math. **54**, 425—505 (1951).

S 4 Skjelbred, T.: Thesis. U.C. Berkeley (1972).

S 5 Smith, P.A.: Fixed points of periodic transformations. Appendix B in Lefschetz, Algebraic topology, 1942.

S 6 Smith, P.A.: Fixed point theorems for periodic transformations. Amer. J. Math. **63**, 1—8 (1941).

S 7 Smith, P. A.: The topology of transformation groups. Bull. Amer. Math. Soc. **44**, 497—514 (1938).

S 8 Smith, P. A.: Transformations of finite period II. Ann. of Math. **40**, 690—711 (1939).

S 9 Smith, P. A.: New results and old problems in finite transformation groups. Bull. Amer. Math. Soc. **66**, 401—415 (1960).

S 10 Spanier, E. H.: Algebraic Topology. New York: McGraw-Hill 1966.

S 11 Steenrod, H.: The Topology of Fibre Bundles. Princeton: Princeton University Press 1951.

S 12 Steenrod, N.E., Epstein, D.B.A.: Cohomology Operations. Ann. of Math. Studies **50**, (1962).

S 13 Su, J. C.: Periodic transformations on the product of two spheres. Trans. Amer. Math. Soc. **112**, 369—380 (1964).

S 14 Su, J. C.: Transformation groups on cohomology projective spaces. Trans. Amer. Math. Soc. **106**, 305—318 (1963).

S 15 Swan, R. G.: A new method in fixed point theory. Comm. Math. Helv. **34**, 1—16 (1960).

S 16 Stewart, T. E.: Lifting group actions in fibre bundles. Ann. of Math. **74**, 192—198 (1961).

T 1 Tomter, P.: Transformation groups on cohomology product of spheres. Invent. Math. **23**, 79—88 (1974).

W 1 Wang, H. C.: Two-point homogeneous spaces. Ann. of Math. **55**, 177—191 (1952).

W 2 Wang, H. C.: Homogeneous spaces with non-vanishing Euler characteristics. Ann. of Math. **50**, 925—953 (1949).

W 3 Wang, H. C.: Compact transformation groups of S^n with an $(n-1)$-dimensional orbit. Amer. J. Math. **82**, 698—748 (1960).

W 4 Weil, A.: L'Integration dans les Groupes Topologiques et ses Applications. Paris, 1951.

W 5 Weyl, H.: Classical Groups. Princeton: Princeton University Press 1939.

W 6 Weyl, H.: Theorie der Darstellung kontinuierlicher halbeinfacher Gruppen durch lineare Transformationen. I, Math. Z. **23**, 271—309 (1925); II, Math. Z. **24**, 328—395 (1926).

Z 1 Zariski, O. and Samuel, P.: Commutative algebra I, II. New York: Van Nostrand.

Subject Index

Ergebnisse der Mathematik und ihrer Grenzgebiete

1. Bachmann: Transfinite Zahlen
2. Miranda: Partial Differential Equations of Elliptic Type
4. Samuel: Méthodes d'algèbre abstraite en géométrie algébrique
5. Dieudonné: La géométrie des groupes classiques
7. Ostmann: Additive Zahlentheorie. 1. Teil: Allgemeine Untersuchungen
8. Wittich: Neuere Untersuchungen über eindeutige analytische Funktionen
10. Suzuki: Structure of a Group and the Structure of its Lattice of Subgroups. Second edition in preparation
11. Ostmann: Additive Zahlentheorie. 2. Teil: Spezielle Zahlenmengen
13. Segre: Some Properties of Differentiable Varieties and Transformations
14. Coxeter/Moser: Generators and Relations for Discrete Groups
15. Zeller/Beckmann: Theorie der Limitierungsverfahren
16. Cesari: Asymptotic Behavior and Stability Problems in Ordinary Differential Equations
17. Severi: Il teorema di Riemann-Roch per curve, superficie e varietà questioni collegate
18. Jenkins: Univalent Functions and Conformal Mapping
19. Boas/Buck: Polynomial Expansions of Analytic Functions
20. Bruck: A Survey of Binary Systems
21. Day: Normed Linear Spaces
23. Bergmann: Integral Operators in the Theory of Linear Partial Differential Equations
25. Sikorski: Boolean Algebras
26. Künzi: Quasikonforme Abbildungen
27. Schatten: Norm Ideals of Completely Continuous Operators
30. Beckenbach/Bellman: Inequalities
31. Wolfowitz: Coding Theorems of Information Theory
32. Constantinescu/Cornea: Ideale Ränder Riemannscher Flächen
33. Conner/Floyd: Differentiable Periodic Maps
34. Mumford: Geometric Invariant Theory
35. Gabriel/Zisman: Calculus of Fractions and Homotopy Theory
36. Putnam: Commutation Properties of Hilbert Space Operators and Related Topics
37. Neumann: Varieties of Groups
38. Boas: Integrability Theorems for Trigonometric Transforms
39. Sz.-Nagy: Spektraldarstellung linearer Transformationen des Hilbertschen Raumes
40. Seligman: Modular Lie Algebras
41. Deuring: Algebren
42. Schütte: Vollständige Systeme modaler und intuitionistischer Logik
43. Smullyan: First-Order Logic
44. Dembowski: Finite Geometries
45. Linnik: Ergodic Properties of Algebraic Fields
46. Krull: Idealtheorie
47. Nachbin: Topology on Spaces of Holomorphic Mappings
48. A. Ionescu Tulcea/C. Ionescu Tulcea: Topics in the Theory of Lifting
49. Hayes/Pauc: Derivation and Martingales
50. Kahane: Séries de Fourier absolument convergentes
51. Behnke/Thullen: Theorie der Funktionen mehrerer komplexer Veränderlichen
52. Wilf: Finite Sections of Some Classical Inequalities
53. Ramis: Sous-ensembles analytiques d'une variété banachique complexe
54. Busemann: Recent Synthetic Differential Geometry
55. Walter: Differential and Integral Inequalities
56. Monna: Analyse non-archimédienne
57. Alfsen: Compact Convex Sets and Boundary Integrals
58. Greco/Salmon: Topics in $\mathfrak{m}$-Adic Topologies
59. López de Medrano: Involutions on Manifolds
60. Sakai: C*-Algebras and W*-Algebras
61. Zariski: Algebraic Surfaces